Water: More or Less

An intimate vision of the world of water as seen through the eyes of a water journalist, an artist, and through the diverse voices of water decision-makers.

21 Diverse Voices

Diverse water policy decision-makers contribute personal essays for this book, bringing additional insight to options for the future of water, telling why they continue to have passion for this issue.

Rita Schmidt Sudman

Rita Schmidt Sudman, Emmy award winning journalist and former executive director of the Water Education Foundation, follows the history of water conflict in each geographic area from Mount Shasta to the Salton Sea. She examines many pressures that set the stage for disagreements and decisions that shaped water policy over decades. She provides context for the past, adds to an understanding of the present and suggests solutions for the future.

Stephanie Taylor

Stephanie Taylor has a dream career making paintings and sculpture for private and public collections all over the United States and in Paris. Combining her love of history and location, she's also a newspaper Opinion contributor. Her paintings, photography and lyrical essays capture images of people and places and describe a slice of the daily lives of fishermen, farmers, biologists and others. She invites the reader along as she hikes the land, traces the path of waterways and talks with those who rely on water for their livelihood. She has created the original art and most of the photography for this book.

Library of Congress Cataloging-in-Publication Data is available for this publication.

ISBN 978-0-9972382-1-1

Book and cover design by Stephanie Taylor
stephanietaylorart.com

Design assistance by Broken Donkey Creative

Editorial assistance by Cindy Nickles

Pentimento Press

a small independent art & architecture press
pentimentopress.com

Independent Book Publishers Association

Second Edition

Mount Shasta
Lake Shasta
Lake Oroville
Folsom Lake
Sacramento
Delta
San Francisco Bay Area
Mono Lake
Hetch Hetchy
CA Aqueduct
San Luis

Graphic inspired by DWR and Water Education Foundation maps

Guest Authors
21 diverse voices

Tom Birmingham
General Manager, Westlands Water District

Jennifer Bowles
Executive Director, Water Education Foundation

Celeste Cantú
General Manager, Santa Ana Watershed Project Authority

Michael Cohen
Senior Associate, Pacific Institute

Kim Delfino
California Program Director, Defenders of Wildlife

Laurel Firestone
Co-Executive Director and Co-Founder, Community Water Center

Letitia Grenier
Program Director and Senior Scientist, San Francisco Estuary Institute

David Guy
President, Northern California Water Association

Thomas Harter
Professor Cooperative Extension Specialist, Groundwater Hydrology,
University of California, Davis

Carl Hauge
Chief Hydrogeologist, California Department of Water Resources (retired)

Campbell Ingram
Executive Officer, Delta Conservancy

Kevin Kelley
General Manager, Imperial Irrigation District

Jeffrey Kightlinger
General Manager, Metropolitan Water District of Southern California

Sunne Wright McPeak
President, Delta Vision Foundation and CEO, California Emerging Technology Fund

David Orth
Member, California Water Commission and former General Manager,
Kings River Conservation District

Tim Quinn
Executive Director, Association of California Water Agencies

Tim Sloane
Executive Director, Pacific Coast Federation of Fishermen's Associations and
Institute for Fisheries Resources

Lester Snow
Executive Director, Water Foundation

Frances Spivy-Weber
Board Member, State Water Resources Control Board

Maureen Stapleton
General Manager, San Diego County Water Authority

Kevin Starr
State Librarian Emeritus and author of the *Americans and
the California Dream* series

Introduction

Sacramento Valley

The Coast

The Delta

Sierra Nevada

The Southland

Imperial Valley and Salton Sea

Some Solutions

Index

Foreword

Rita Schmidt Sudman
Former Executive Director
Water Education Foundation

I have been on a journey - sometimes unknown to us - but bound by the struggle over water.

Water issues have dominated my life in California, an unpredictable and vulnerable state where there is no such thing as "normal." Co-authoring this intriguing book has been a welcome challenge, to synthesize and share 35 accumulated years of observations and experience in a search for options and solutions.

In 1979, after an epic drought hit California, we all woke up to the importance of water in our lives. I eagerly accepted the position of Executive Director of the nonpartisan Water Education Foundation and dedicated myself, and this new Foundation, to educating the public about water. My background in radio and television enhanced our outreach as we tackled all sides of water issues throughout the Western United States.

I covered water from the perspective of a journalist. The deeper I dug, the more facets I discovered to this complicated subject. When I thought I knew something about water, I learned there was another layer that added to its complexity.

While leading the Foundation, I covered the big fact driven picture and was not often focused on individual stories of people personally affected by the periodic inevitable droughts or floods. So shortly after I retired from the Foundation in 2014, I was delighted to join artist and Sacramento Bee essayist Stephanie Taylor in this unique book project which pairs Stephanie's first-person experiences and dramatic art with my articles that give background and context to California water issues. Together, we search for understanding.

In addition to the two elements that Stephanie and I bring to the book, the third part of our triad is the special guest essays that add dimension to the stories and policies. I was gratified that some of the people I respect most in the water world – people from all sides of the issue – agreed to write original pieces. I asked them to ignore their usual policy arguments and briefly focus on why they remain committed to water issues, what keeps them engaged and where they believe we can find solutions. They delivered.

When I thought I knew something about water, I learned there was another layer that added to its complexity.

Stephanie and I grew up in opposite ends of the state. She was raised in Sacramento in a family deeply rooted in Northern California. Like Sacramento born-writer Joan Didion, her family remembers the rise of the rivers in the flood years and the lack of water in the drought years. She lived alongside the Sacramento and American rivers in her daily life. In high school she vividly remembers a great Valley flood when she saw firsthand fields of drowned cattle.

At about the same time, I was in Vista in San Diego County, the daughter of Camp Pendleton Marine. We lived near the San Luis Rey River – a place of periodic drought, floods and continual fights over water rights between Native American tribes and the federal government. Water united all our families in that little valley. At our mission school, we were children of the Marines, Japanese and Mexican truck farmers, citrus, avocado and poinsettia growers and small town business people. Before the State Water Project was built, we survived with water brought in from the Colorado River and the small local supply that allowed Vista to exist.

I see now that Stephanie and I have been on a journey - sometimes concealed from us - but bound by the struggle over water. Today with climate change a reality and droughts followed by floods, the subject is more important than ever. *Water: More or Less* is our effort to add clarity to the continuing struggle. Please accept our invitation to join us for an exploration into the world of California water.

Rita Schmidt Sudman

Dedication to Joan Didion

Joan Didion

It's easy to forget that the only natural force over which we have any control out here is water, and that only recently.

- Joan Didion

This book is dedicated to writer Joan Didion, a Sacramento and California native whose respect for water is featured in many of her writings and personal discussions. She truly has "water in her blood," perhaps inherited from her pioneer family who journeyed across America in a covered wagon and settled in arid California. Her late husband, screenwriter John Gregory Dunne, understood this part of her, saying that she wrote about California water "in the most reverential, even worshipful, terms."

Joan Didion's description of being a young girl in Sacramento before the big dams controlled the rivers was often one of dread. She listened to the clanging of home water pipes in the drought years and kept watch on the vulnerable levees in flood years. In her 2003 book, *Where I Was From*, she wrote, "I would have to say the rivers are my strongest memory of what the city was to me. They were just infinitely interesting to me. I mean, all of that moving water. I was crazy about the rivers."

Who else but Joan Didion would turn an ordinary visit to the CVP/SWP Joint Operations Center in Sacramento into an inspiring and poetic event? Her essay, "Holy Water" was about that visit and was included in her book, *The White Album*, 1979. In it she writes with fascination of the delivery of vast quantities of water from Oroville Dam to Southern California in eight days.

"As it happens my own reference for water has always taken the form of this constant meditation upon where the water is, of an obsessive interest not in the politics of water but in the waterworks themselves, in the movement of water through aqueducts and siphons and pumps and forebays and afterbays and weirs and drains, in plumbing on the grand scale."

Of that visit she wrote, "I wanted to be the one, that day, who was shining the olives, filing the gardens, and flooding the daylong valleys like the Nile. I want it still." Joan Didion actually makes water deliveries sound fascinating. I get it.

It was only natural that after I began my career in water, I wrote to her, and to my surprise and delight, she wrote back. Our correspondence continued on and off over 34 years.

She read the Water Education Foundation's magazine, Western Water, and encouraged me through the years. When the Water Education Foundation celebrated its 35th anniversary a couple of years ago, she sent me a note saying she regretted not being able to be with me at the Foundation's Sacramento celebration but, "I send all my admiration and thanks for everything you do."

Rita Schmidt Sudman

An Appreciation to Hans Doe

Hans Doe

This book is written with appreciation to the memory of Hans Doe and his able partner and wife, Margaret - two people who had a passion and vision of water for their time and for the future.

Hans was born in Norway and eventually made his way to the small southern California town of Vista where he became an avocado and macadamia nut rancher. Concerned about the reliability of the area's water which came from wells and distant Lake Henshaw, he was first elected to the Vista Irrigation District in 1951 and was a leader in regional water issues for 35 years. Margaret served as his administrative assistant in business and water issues and he often credited her with his success.

He led the local 1954 fight to ensure water reliability for Vista by urging the vote to join Metropolitan Water District of Southern California which would facilitate Vista getting water from the Colorado River. While serving on the water district, Hans also worked to resolve statewide water issues and was an early supporter of the State Water Project.

I grew up in Vista and remember my mother often saying, "God made the earth and Hans Doe made Vista." My parents were grateful to Hans for his efforts to secure a firm water supply and provide flood protection for our little town. I knew Hans in my early career and saw that he was widely respected for keeping his word and working for the benefit of Vistans and other Southern Californians.

By the time Hans died in 1988 at the age of 85, he and Margaret had set up a trust which has since funded a number of educational water programs including scholarships to students from Vista. The Water Education Foundation's beautiful book, *Water and the Shaping of California*, written by Sue McClurg, received support from the Trust in 2000.

Hans was a man of his time and his involvement with water issues then is different from the issues facing Californians today. Still many issues – reliability, cost and quality – remain similar. Hans and Margaret Doe understood the value of water. I often heard Hans quote the old Ben Franklin saying, "When the well is dry, we know the worth of water."

Rita Schmidt Sudman

Preface

Stephanie Taylor
Artist/Writer

Painters have often taught writers how to see.
James Baldwin

If I say that my contributions to this book are special because I'm an artist, that artists provide a particularly intimate and valuable view of the world, it's not enough.

I could start by stating what I am not. Not a water expert, not a scientist, nor historian, lawyer, farmer, fisherman, water manager, and I don't even like to swim.

I could start by listing all the things that give me some credibility to write about water, like being born into a valley of extremes—heat, flood and threats of flood.

I could say I lived on the confluence of two magnificent rivers, spent idyllic summers on the American near the gold discovery site of Coloma. That I vacationed at a tiny lake near Clear Lake, that perfection was a day spent rowing, floating and fishing for inedible perch. I water-skied on the Sacramento, at Tahoe and Folsom lakes, and skied on Sierra snow. I lived four blocks from endless Southland beaches—sunning, biking, jogging, skating—and that my kids were born there, tiny beach babies.

The process of writing and sketching slows down time, time to let perception seep in.

A degree in history, a deep quest for connections, causes and effects in this complicated state, and the earth in my blood as a fourth-generation Northern Californian inspires my entire career as an artist. The role of location and history informs my paintings and sculpture-projects, which continue to be an excellent excuse to learn, with years of research into California culture and history.

Brikstal Glacier
Norway, 2007

My cousin pointed to towering boulders and said that it all starts with water creeping into a tiny crack in a rock. Freezing, expanding, forcing, growing cracks, year after year, until a segment comes crashing down.

After returning to Sacramento, I lived near the American River, with hours spent along the river, with lush fragrances of summer heat, and sharp winter freezes. Always, the sound of water drifting to the sea.

Norway in 2000, I met my grandmother's family, many of whom still live under the very same glaciers as skiers, guides and farmers. Norway in 2007, responding to my cousin's call that the glaciers were melting so fast that they wouldn't last our lifetimes, I followed the water from the top of a glacier down, one drop at a time and came home captivated by water, its past and future.

It's an irony that writing about water is often so dry when the subject itself is so wet.

I spent an entire year making a body of work about water. Paintings inspired by the glaciers, ceramic drops as sculpture, and ceramic buoys that I floated down the American.

One day I sat high on a hill, looking down into Tomales Bay and the San Andreas Fault, wrote a poem, made a sketch, and sent a pitch to Gary Reed, Opinion editor at the Sacramento Bee. Since 2011, my curiosity and passion for water have only grown with my Bee series, "California Sketches."

The art is in the detail.

It's impossible to write about California without touching water in some way. My essays and paintings are about people in a water landscape. It's a natural step to take—to gather essays, to want to learn more, to search for meaning in this complex chaos. To reconnect with Rita, Executive Director of the Water Education Foundation, and add contributions by others who know a lot more about water than I.

And thus, this book.

"240 Tiny Paintings," ceramic,steel, plexi, 72" x 18", 2008

Introduction

Drought in California: A National Issue

Kevin Starr
State Librarian Emeritus and author of the *Americans and the California Dream* series

Many were predicting that California was over, finished, hasta la vista! Not so. Not for now at least.

The latest drought is offering California one final opportunity to reinvent itself – in part through reformed water usage – and thus be able to maintain its viability as a nation state.

This drama of reinvention in response to environmental change and scarcity is of significance not only to California but to the rest of the nation and major portions of the planet as well.

California, the late novelist and historian Wallace Stegner once observed, is like the rest of the United States – only more so. In no case is Stegner's citation of the representative nature of the California experience more relevant than in California's current effort to deal with serious drought conditions now four years in duration.

Each region of the United States, first of all, must face its own distinctive challenge from Mother Nature. Wintry New England, as we have recently experienced, must deal with snowfalls that brought that region to a near total halt. Sudden and unexpected hurricanes and tornados devastate the southeast and Great Plains Midwest with regularity. The Mississippi has a tendency toward catastrophic flooding. In the 1870s Kansas was twice denuded by voracious hordes of locusts. The following decade, in May of 1889, a ferocious rainstorm destroyed the dam above Johnstown, Pennsylvania, that took some 2,200 lives.

Admittedly, California has more than its fair share of these environmental vulnerabilities. Its coastal regions are crisscrossed by major fault lines that periodically slip and grate against each other to cause earthquakes that, in the case of the April 1906 San Francisco quake, set in motion natural forces and human decisions (dynamiting of buildings, for example) that destroyed the city and did serious damage to its hinterlands. Since time immemorial, periodic wildfires have with regularity cleansed the chaparral and woodlands of the California coast. Of late, these periodic conflagrations have scorched to the earth advancing suburbs, driving thousands into homelessness and financial ruin.

So then: each American region faces its own environmental vulnerabilities. California only more so. In the case of the current drought, however, it is perhaps difficult for other American regions – especially for those for whom snow, rainfall and flooding are frequent occurrences – to see in such water scarcity a prophetic example of the difficulties the United States might face in a water-scarce future.

Citizens of the Far West and Southwest, by contrast, have long since learned to live in places where water was scarce. One hundred and thirty-seven years ago, in Lands of the Arid Region (1878), the geologist-explorer John Wesley Powell warned a nation on the verge of settling and developing the Far West that lands beyond the 100th meridian were water-scarce, arid to semi-arid even, and would require water engineering and prudent use patterns to sustain significant populations. In the case of the desert regions of the Southwest, moreover, extending from Southern California across what is today called the Sunbelt, the challenge was even more formidable. Indeed, in earlier times, maps depicted the unsettled Far West as the Great American Desert, conferring on that region a level of aridity that seemed to preclude any settlement whatsoever.

The latest drought is offering California one final opportunity to reinvent itself – in part through reformed water usage – and thus be able to maintain its viability as a nation state.

In Southern California, into which the deserts of the Great Southwest either originated or terminated, depending upon one's coastal orientation, it took nearly 100 years for Americans to learn to live successfully in these environments. Visiting Los Angeles in the early 1920s en route to New Mexico, novelist D. H. Lawrence noted the pervasive aridity of Southern California and stated that he would not be surprised that were he to return to Los Angeles a few months hence, he might very well find the city abandoned and tumbleweeds running down its streets.

From the Owens Valley, the Los Angeles Aqueduct cascades into the San Fernando Valley

Almost 100 years out from Lawrence's observation, Californians are coping with the fact that, yes, water is indeed in short supply, and drought reappears with documented regularity. Those 100 years witnessed the creation by local, state and federal agencies of a water delivery system unparalleled in the history of the human race. But it took more than a water delivery system to service nearly 40 million residents. It took water, and water was frequently absent in the annals of American California, starting with the great drought of the mid-1860s that wiped out the cattle industry of the southern counties.

Folsom Lake in the drought of 2015

By the early 1900s, thanks to the pumping of groundwater made necessary by water scarcity and periodic drought, Southern California was facing a crisis of subsidence as land levels, pumped dry, sunk dangerously throughout the region. It was at that point, however, that both Los Angeles and San Francisco metropolitanized themselves through dam and aqueduct technologies that brought both water and hydroelectricity to these water scarce regions.

In 1910, California had a population of 2,377,549. Over the next 100 years, in significant measure through water engineering and hydroelectricity projects, California invented itself as a mega nation state capable of supporting 38 million residents and looking toward 50 million by the mid-21st century.

The latest drought is offering California one final opportunity to reinvent itself – in part through reformed water usage – and thus be able to maintain its viability as a nation state. Perhaps the drought is giving California its final chance to heed the advice offered the Far West by Powell. This was advice that did not include the massive importation of the lawn-centered English garden or golf course, or backyard swimming pools, or 20-minute hot showers or running the faucet while brushing your teeth. But that's the easy part – cutting back by 25 percent on residential usage, as decreed by Governor Jerry Brown. Ahead lie more demanding decisions regarding agricultural usage, as well as the more challenging issues of desalinization, pricing, increased storage capacity, allocation of water rights, planning, architectural, and landscape adjustments, and crucial to it all: a change from a culture of consumption based on abundance to a culture of prudent use based upon scarcity.

This drama of reinvention in response to environmental change and scarcity is of significance not only to California but to the rest of the nation and major portions of the planet as well.

This drama of reinvention in response to environmental change and scarcity is of significance not only to California but to the rest of the nation and major portions of the planet as well. Let's start with major portions of the planet. Water scarcity will inevitably be part of the global future in the age of global warming. Already, vast portions of sub-Saharan Africa are suffering from the horrors of long-term drought.

But so too do the arid regions of the American Southwest have their own vested interest in the current California story. They share the same vulnerability and the chances are that they might share similar water-scarce futures.

And for the rest of the nation, the challenge might not involve water – unless in the case of devastating floods! – but it will involve related environmental challenges in the century now unfolding: challenges already present – snowstorms, hurricanes, tornados, floods – or challenges to come in a globally warming era.

In times past, faced with near-catastrophic challenges, natural or man-made – earthquakes, fires, floods, drought, a recent near-collapse of its electrical system, a recent near-collapse of its state finances – many were predicting that California was over, finished, hasta la vista! Not so. Not for now at least. Not while nothing less than the nation itself depends in part upon the outcome.

Kevin Starr

American River near Sacramento

Interview

A Journey Towards Solutions

Lester Snow
Policy Advisor,
Water Foundation

There's a lot of attention on much more holistic, integrated ways of looking at water
-Lester Snow

Lester Snow, the mastermind behind countless water resources management projects, has been involved in both public and private sector water issues, and on regional, state and federal levels of government. Snow served from 1988-1995 as the general manager of the San Diego County Water Authority after leaving the Arizona Department of Water Resources. From 1995-1999, he was the executive director of the CALFED Bay-Delta Program, which included a team of both federal and state agencies. In 1999, Interior Secretary Bruce Babbitt appointed him as the Mid-Pacific regional director of the U.S. Bureau of Reclamation. In 2001, he went into the private sector as a consultant until 2004, when Governor Arnold Schwarzenegger appointed him as director of the California Department of Water Resources. In 2010, Schwarzenegger appointed him as secretary of the California Natural Resources Agency. Currently he is the executive director of the Water Foundation and a member of the California Water Service Group Board of Directors and the Water Education Foundation Board of Directors.

Rita Schmidt Sudman interviewed him during the summer of 2015.

Sudman: You are the only person I know who's held every job in California water. Take me back to when you first came to this state from Arizona.

Snow: I think probably the first thing that hit me was the sheer magnitude of the water system and the incredible fragmentation of water management. So many agencies, so many different jurisdictions that had impact on it—impressive in one sense in terms of the historical reliability of the water supply. But also daunting in terms of how you ever get change in California when so many people are involved.

Lester Snow and Rita Schmidt Sudman

Sudman: You were managing water in San Diego, a place that was relying almost totally on imported water.

Snow: At that time 90% to 95% of the water in San Diego County was imported water, which meant we were very vulnerable to things that happened on the Colorado River and in Northern California and so our destiny was not our own. You had to pay attention to policy issues and conflicts in other parts of the state.

Sudman: After a short while, you left San Diego and ended up in the thick of it in Sacramento.

Snow: Being in San Diego, when you look at where your vulnerability is, it drew me into the State system right away. Follow the stressors; follow the problems that might affect your water supply. And the more I engaged, the more I became part of the discussion on what to do about long-term reliability and sustainability. Then all of a sudden you're in the Delta.

CALFED partially funded the intertie between the state and federal projects

Sudman: That's when you took the job running CALFED.

Snow: When I was in San Diego, I became quite engaged in something called the Three Way discussions, which was an initiated conversation between urban, ag users and a subset of the environmental community. We were all trying to avoid conflict and a showdown in the future and to work out how we balance the system so that ag and urban have a supply and there's a healthy ecosystem.

Sudman: I remember covering that process in about 1990. It was the first time that those parties really sat down in good faith and started talking about the hard issues.

Snow: For me, relatively new to California, it was just a great way to understand the issues and understand perspectives.

Sudman: And did that process actually lead to any accomplishments?

Snow: It did. One of the problems with California water is that the accomplishments are never quite as grand as you hope they will be. But that effort led to something called the Bay Delta Accord and so it averted a showdown between the Environmental Protection Agency (EPA) and water users on how much water could be diverted out of the Delta.

Sudman: That process lead to avoiding a state and federal conflict.

Snow: That's a good way to put it. So there was agreement on the amount of flows in the system and the concept of investing in habitat restoration and limiting diversions at certain times of the year. It was a template that's still in play. It kept peace for a number of years.

Sudman: Did the Accord lead directly to the federal-state CALFED process?

Snow: It did, and one of the provisions in the Accord was that there would be an official state-federal process to develop a long-term plan for the Delta.

Sudman: Who was funding it?

Snow: When the Accord came together in 1994, it was a state-only process and the agreement was that there had to be a state-federal process. And so that was set up in early 1995 and they were looking for somebody that could work with and for Governor (Pete) Wilson and with and for Secretary (Bruce) Babbitt, the two leaders of the process. I got to know both parties from working first in Arizona and then in San Diego. I decided to jump into the Delta fully.

Sudman: In this case, you were able to work with both a Republican California governor and a Democrat federal Interior Secretary.

Snow: To the extent that they had differences of opinion, those opinions didn't really fall along political lines.

Sudman: What did the CALFED process accomplish and why did it fade away?

The information that was developed in CALFED is constantly undervalued.
-Lester Snow

Snow: I think the information that was developed in CALFED is constantly undervalued. There was a lot of work done on the ecosystem and on habitat issues and a more integrated approach to water management.

Sudman: That was new.

Snow: Yes, and that wasn't part of the equation before. Probably the most significant deviation from the way it was done in the past was having a biologist be part of certain operations decisions. For example, when to curtail pumping when the fish are in what location? And when can you start pumping again? And that was a major innovation in terms of collaborative management of a large system. But in terms of moving forward, the concept and the final Record of Decision was that moving water around the Delta is clearly the most effective way. We can try other projects and continue to move water through the Delta and combine that with other kinds of investments. And if that fails, then let's move on to some sort of isolated conveyance. I think the biggest issue (diminishing CALFED) was the changes in administrations. A deal came together under the Clinton administration, under the watchful eye of Secretary Babbitt and then it really was finalized just as the Bush administration came in. So you didn't have the same ownership.

Sudman: Is there anything left of CALFED today?

Snow: Some things that got started like science panels and science coordination. Some of the habitat restoration that was identified is proceeding. It made some changes just in the way things were done, but never got the brass ring.

Sudman: There have been so many Delta processes through the years, including the latest makeover of the Bay Delta Conservation Plan, now called California Water Fix. Haven't you and I seen so many Delta commissions, task forces, strategic plans, scientific studies and appointed panels? Do they lead to solutions or are they dressed up groups that just fade away when another governor comes into office?

Snow: All of the above. The trail of efforts that haven't led to the big solution is an indication of two things—the consistent importance of the Delta and the recognition it needs to be fixed. So people keep going back to it even though there's no political upside of taking on the Delta issue because they see the importance of the Delta and the fact that the status quo will lead to a disaster one day. But the many ships on the rocks that have crashed in this effort is just an indication of how extremely difficult it is to forge a solution that will stick. We've been struggling with the structure of the Delta for about 40 years.

The trail of efforts that haven't led to the big solution is an indication of two things—the consistent importance of the Delta and the recognition it needs to be fixed.
-Lester Snow

Sudman: There are many physical fixes discussed. We all know about the big fight in 1982 over the Peripheral Canal proposal to carry water around the Delta. There have been other Delta solutions proposed including the tunnels. What do you support?

Snow: I firmly believe that the status quo on the Delta will fail us. It's a matter of time. And we're seeing a steady deterioration of the environment. We continue to see Delta islands subside. If there's an earthquake or if we lose enough species, action will have to be taken and that's in contrast to implementing a well-thought out strategy to move forward. So I think that the conveyance and ecosystem issues have to be dealt with. I don't think it's an option. What I don't happen to have an opinion about is the two tunnels, and what does the output need to be?

The biggest issue that seldom gets addressed is the governance issue. If you build a facility, who's in charge?

Sudman: There has always been an issue of assurances and trust. In any agreement, if people don't trust each other, the agreement isn't worth very much.

Snow: That trust issue gets overstated in some cases. Meaning you could run the pumps full-time now but we don't because of the rules that govern the usage of those pumps. So, there are laws and regulatory structures that regulate how they can be used. But if that's the key issue, then let's get on to the key issue and talk about an alternative governance structure—and that doesn't seem to come up very much.

Sudman: Why?

Snow: I think there's some interests that don't want the facilities built regardless of who operates them. I think there are some that have no interest whatsoever in seeing a balanced solution.

The California Delta

Sudman: Would you be referring to some people in the environmental movement and the agricultural interests in the Delta?

Snow: I think that's fair. It is so extremely complicated and there are different periods in time when people have decided no, we're not going to build that facility. It goes all the way back to the Central Valley Project. The original plan included a conveyance facility in the Delta. And then in the mid-1970s, in Governor Reagan's administration, they decided against it because the federal government didn't want to pay for it.

Sudman: The current governor supports the tunnels. Do you see hope in his leadership?

Snow: I do and for anything to happen, you're going to need gubernatorial leadership. This isn't something that can happen at a lower level. I think eventually there needs to be California business community engagement and labor and environmental community engagement. The other side of this is there has to be assurance that habitats are going to be restored. And all that's still in play. You can't make any progress without the governor being in charge.

Sudman: Do you think Governor Jerry Brown has water in his blood—from his dad?

Snow: One could assume that, yes.

Sudman: In the light of the lack of progress in a Delta solution, Southern California has come to the realization that they need to diversify their water portfolio. Do you think we will see a time when the Delta gets abandoned as the transfer system for water because we just can't reach a solution?

Snow: I don't think that will ever happen. I do think, consistent with the Delta Act that was passed in 2009, there has to be a concerted effort to reduce dependence on the Delta. But I think that the Delta can and probably always will provide a reliable portion of water supply for the San Joaquin Valley and Southern California. I don't know what the right amount is, and I think it's less than historic deliveries out of the Delta, but it's I think an integral part of long-term water supply. But developing (a Delta project) as an excuse for not doing diversification is just unacceptable. One very sore point with me is that here we are in a drought and we're discharging over a million acre-feet of wastewater into the ocean. That's the kind of stuff that just has to stop.

Sudman: Do you think this is a crisis that's too good to waste? You mention using water runoff from our yards rather than funneling it into the gutters.

Snow: Yes, I think there's a lot of attention on much more holistic, integrated ways of looking at water. Actually Mayor (Eric) Garcetti in Los Angeles has outlined a "one water" approach that is about capturing the water that falls on properties—residential, commercial, developing the stormwater, expanding wastewater recycling and conservation. That's what we need to see everywhere. That's a supply that has to grow over time. It just becomes a part of your portfolio.

Sudman: That takes investment. How important is it that we make these investments and that people understand it will cost them more?

> *Water in most of the state is probably our cheapest utility, far cheaper than cable and phone.*
> *-Lester Snow*

Snow: I think when you put it in terms of taking advantage of the attention that the drought provides, I think that's really important. Water in most of the state is still probably our cheapest utility, far cheaper than cable and phone. We are victims of the past investments where people 60 to 100 years ago invested in these large scale systems and we got used to a high level of reliability. It really is time to reinvest in this more diversified approach.

Sudman: Is it important for people also to realize that as they use more water, they should pay more?

Snow: Absolutely. Every utility needs to have tiered pricing.

Sudman: Now let's turn to water rights. We have a system in California of dual water rights: riparian and appropriative. Earlier in our history, the courts always ruled for the riparian owners. Finally the voters in California passed a constitutional amendment in 1928 that said water belongs to all the people of California and it shall only be used for beneficial and reasonable uses. Now we are again in a time of questioning established water rights against societal uses. Do you support any kind of redistribution of California water rights?

Snow: Not really. And what I mean by that is that we have an arcane and complex water rights system. In addition to riparian rights, there are pre-1914 and post-1914 rights and it gets pretty complicated. But I think the main issue is we need a better market system and to have a better market system, we need a better data system. In California, despite the fact we're the world leader in information technology, our water rights exist in paper files at the State (Water Resources Control) Board, not in online digital files. And we need to change that so that there's a better markup way for people to move water on a willing seller and buyer basis. Then I think we avoid a lot of this conflict. Now, what I would add is if we get into the 10th year of the drought, people are going be expecting the State Board to exercise its responsibility under waste and unreasonable use and the Public Trust Doctrine. That will be temporary. But I think a water rights system is something that can smooth over some of these allegation issues in a much more logical and non-conflict way.

Sudman: Water marketing with willing sellers and buyers has been slow to develop.

Snow: There's some form of marketing that goes on just about every year, but during the drought, most of the transfers are ag to ag. It's people who have big investments in orchards that can't afford to dry it up for a year. That's not the way an orchard works. But that system right now is very opaque. It's hard to tell who is moving water. It needs to be much more transparent. There needs to be much better data on who has rights, who's exercising their rights, how much they are using, and that's hard to know right now.

There needs to be much better data on who has water rights, who's exercising their rights, how much they're using, and that's hard to know right now.
-Lester Snow

Sudman: Who will make that happen?

Snow: There's starting to be that movement of collecting more information and perhaps with less controversy because I think water right holders see how that it might be in their best interest to know what everybody else is using and to understand if they're being impacted by someone else.

Sudman: In recent years, you've concentrated a lot on groundwater issues, promoting a change in groundwater law and regulation. Why has that become so important to you?

Snow: You can't have sustainable water management if you have one water supply that you don't even measure and you just use as much as you want. It's like the hole in the bucket of sustainability and that is classic deficit spending. You can't balance your budget if you never know how much is in your bank account.

Photo: CA Department of Water Resources

Kern Water Bank

Sudman: This groundwater bill was fought by agricultural interests until some started seeing that groundwater was being taken from them by other users. Do you think that this drought made it possible to pass a state groundwater law?

Snow: Yes. There are a lot of things that happened last year that enabled this to pass. But one of them was that some in the ag community stepped forward saying enough is enough. We had some people that saw that their long-term sustainability and viability was really based on effective groundwater management and I don't think we would have gotten the bill absent that.

Sudman: Do you think, even with this groundwater law, we have enough time to save this resource?

Snow: Well, I do. I know people are concerned about that, but you can't bring it into balance immediately unless you destroy the economy. And in some areas, they don't even have enough information to say how much you should be pumping. So it's going to take time to work through this. We've seen regions that are trying to jump on this issue and get started. And I have yet to see a region whose plan it is to not take any action until the last deadline.

Sudman: What else motivates you in water policy these days?

Snow: It's all about sustainability. Water is one of those hybrid things. When you divert it out of the stream, you treat it like a commodity and when it's in the stream, it's an essential part of the natural environment. And so if we can't figure out how to manage our water sustainably in a time of climate change, then we're going to have all kinds of societal problems and food supply problems worldwide. It's an important issue and that's why at times, I probably appear to be critical of agriculture, especially historically around groundwater, but I'm a big defender of California agriculture. The key for a highly productive and globally competitive California agriculture is achieving that level of sustainability. You cannot deplete your groundwater and be competitive.

You can't deplete your groundwater and be competitive.
-Lester Snow

Sudman: What else do you want to accomplish before the end of your career?

Snow: The big thing for me is trying to get some form of a fee system in place. Right now we make these episodic investments when we can pass a water bond. And that's great and that stimulates a certain level of investment, but I happen to believe we need a steady revenue stream that would come from a fee. Another issue that's really unaddressed is we have all of these disadvantaged communities that don't have adequate drinking water. You can get the capital money to build a treatment plant (for the disadvantaged community), but then they have to operate it. And that means that their rates are going up, and they're a disadvantaged community, so who pays for the operation and maintenance?

Also in restoring habitat, we can find the capital money often, but it's a question of who maintains the habitat and how do you pay for that? So, there are a lot of issues that I think means that the state at some point needs a fee system, whether it's called a public goods charge or something else, that brings revenue in to invest in long-term sustainability.

Every utility needs to have tiered pricing.
-Lester Snow

Sudman: How is that going to be accomplished?

Snow: It's going to take courageous political leadership and so it's probably something that can only be done during a drought where people are much more attuned to the problems or the weaknesses in our water management system.

Sudman: Water has been your life's work and mine. Would you encourage younger people to continue where we leave off? Do you think there's a good future for them?

Snow: Absolutely. Water's going to do nothing but become more important in societies all across the world.

Canoe on the Cosumnes River

Sacramento Valley

Rice

Farming

Sacramento River

Levees

Wetlands

Ducks

Floods

Shasta Dam

Water Rights

Acre-Feet

Land Boom

Salmon

Sacramento Valley

The Three Shastas

Stephanie Taylor

Snow melts and begins a migration into the cone of this ancient volcano, trickling and filtering through fragments of lava.

One fine spring day in 1875, John Muir hiked to the summit of Mount Shasta. In awe, he watched the weather change. “Storm clouds on the mountains - how truly beautiful they are! - floating fountains bearing water for every well; the angels of streams and lakes.” He was surprised, unprepared for the blizzard that followed, and spent a tortuous night surviving between snow and hot sulfuric springs in “the pains of a Scandinavian hell, at once frozen and burned.”

Muir had hiked to the place water begins. Snow melts and begins a migration into the cone of the ancient volcano, trickling and filtering through fragments of lava.

Fifty years later, the snowmelt emerges from the mountain at a park in the City of Mount Shasta, as the headwaters of the Sacramento River, beginning a 447-mile journey as it flows toward the sea.

Pristine water gushes from three openings in the rock and pauses in a small pond. At its edge, I watched a fuzzy caterpillar arrive, a symbol of metamorphosis—like the water that has transformed this state from a semi-arid desert into our bountiful Central Valley.

Sixty miles downstream, the river will merge with three others, each contributing to the largest reservoir in California—Shasta Lake.

Wedged into a ravine on the southwest side of the lake, Shasta Dam captures the precipitation and distributes the water. It’s a massive structure, 60 stories high—a feat of human ingenuity and enterprise. Constructed in the late 1930s and completed in 1945, the dam is managed by the U.S. Bureau of Reclamation.

The primary purpose of Shasta Dam is flood control and water storage. Irrigation is second, often in conflict with the myriad needs of environment, municipalities, farmers, fishermen and power. Massive pipes transport water from the lake to five huge turbines in the powerhouse, an enormous white space saturated with the roar of electrical generation during periods of peak demand.

From the 30-foot wide curving road on top of the dam looking north, the surface of the nearly full lake appears somewhere between green, blue and vast. Looking south down the spillway to where the Sacramento River continues, perpetually seeping water creates a palette of vertical striations of moss, wet and dry, from chartreuse and orange to white to black.

Larger volumes of water are released in response to water temperatures and salinity levels in the Sacramento-San Joaquin Delta. A collaboration of state and federal agencies strive for a balanced system and flow for salmon and other aquatic species. All stakeholders in the diverse watersheds play a critically delicate and contentious dance.

We heed the words of Muir at Mount Shasta, appreciating the blessings and perils that rain clouds bestow. His voice is a mindful murmur of fragile resources as the river flows south.

Fighting Floods and Politics

Rita Schmidt Sudman

The history of the Valley is a story of how prevailing social movements shaped the land and people of the Valley.

California's Sacramento Valley today is a far different place than the early pioneers found after the arduous journey to this promised land. Their struggle to bring water to grow crops while avoiding terrible floods helps us understand the vast Central Valley and its people. The history of the Sacramento Valley is also the story of how prevailing social movements shaped the land and people of the Valley.

After the influx of pioneers during the Gold Rush years, land was cleared for farming. Wheat became the Valley's dominant crop since it did not require any irrigation or refrigeration to bring it to market in the East. With the arrival of the railroad's new refrigerated cars, Valley-grown fruit could be chilled and preserved during the journey east. As fruit became a major crop, wheat prices dropped and wheat farming diminished. Fruit required irrigation. If the land had access to Sacramento River water or if the farmers could dig productive wells, land increased in value.

Early Battles

But for the historic cycles of drought and flood that devastated the Valley, things might have worked out for the early farmers. They had been warned by the native people that great floods periodically covered the entire Sacramento Valley, and were often followed by periods of devastating drought in which animals died and plants withered. In fact, floods from the Sierra Nevada watersheds often created a huge inland sea covering the entire Valley in the winter months. Captain John C. Fremont actually camped on the Sutter Buttes in the winter of 1846 as it was the only land not under water.

Floods also had benefits. During overflow periods, woodlands of oak, cottonwood, willow and ash lined the river banks up to five miles wide, creating a nutrient rich and protective environment for salmon and other creatures.

The Sacramento River

But to the settlers fishing wasn't important. Farming was their livelihood. After statehood, Congress passed the 1855 Swamp and Overflow Act to encourage American farmers to drain and "reclaim" river marsh lands, turning them into productive farmland. The cost of $1 for an acre of land would be refunded to the purchaser when the land was transformed into productive farmland. At first, there was a 320-acreage limit to qualify, but in 1868 that limit was dropped. The land boom was on - with one catch; it was up to the buyer to build the levees to protect their lands from the dreaded floods.

Thus began a time historians call the Laissez-faire water period, a time in which individuals—not governments—had the authority and responsibility for flood control and water impoundment infrastructure. Several attempts were made to give the state some authority over the ensuing levee construction confusion through the creation of swampland districts. These attempts failed, although some swampland districts became the reclamation districts of today.

As the land boom continued, inexpensive wetland marshes were snapped up by speculators. By 1871, almost a half-million acres of these overflow lands in the Sacramento Valley were owned by just 30 people.

To claim the land, the prospective owner had to prove that the land was wet most of the year. Abuses were notorious. For example, one speculator known locally as the "admiral," supposedly secured his land by testifying he traveled over the land in a boat, leaving out the fact that the boat was sitting atop a horse-drawn wagon!

Hydraulic mining fight

For farmers in this time, periodic flooding was aggravated by the debris from hydraulic mining practices. Huge mechanized water hoses destroyed mountainsides in an industrial-scale extraction of gold. Farmers - and the railroads that carried their produce—fought the powerful mining interests for years to stop this destructive practice. They succeeded in 1884 when a federal court essentially outlawed hydraulic mining. Even so, the remains of this destructive practice adversely affects Valley water quality to this day.

Also around this time, in another effort to unify the development of water projects, the state legislature enacted the 1887 Wright Act to allow the formation of public irrigation districts, to acquire water rights, construct water projects, sell bonds and tax property. It was an important start toward a unified approach to building water projects, though many private water projects continued with mixed results.

The 1919 Marshall Plan laid out a more comprehensive flood and irrigation plan. The public now supported unified water solutions that would involve the state and federal governments.

As the irrigation movement grew, California became part of a new social movement sweeping several states as Progressives gained power. In California, The Progressive Party's Governor Hiram Johnson believed in organized, large-scale water development, and ordered the gathering of information and data for the first state water plan. In a 1914 California election, which was in fact a Progressives-enacted referendum on already passed legislation, the public approved a new water code that asserted state control of water and set up a permit system. The state engineer, William Hammond Hall, had previously devised an integrated flood control and water development plan for the entire Central Valley. Although that plan was not developed during this period, it signaled the end of the Laissez-faire water period. Finally, the 1919 state Marshall Plan laid out a more comprehensive flood and irrigation program. The public now supported unified water solutions that would involve the state and federal governments.

The era of the large water projects was beginning.

When Shasta Dam was built in the 1940s, the river finally was managed. The dream of the early settlers to control and use the water of the Sacramento River was finally realized.

Another big Sacramento Valley water fight in the early 20th century involved support of the U.S. Army Corps of Engineers' plan to narrowly channel the Sacramento River for flood control, boat navigation, and to scour the remaining hydraulic sediment. The opposing idea was to allow some breathing room for the river by creating a bypass system – essentially parking water off the river to relieve flooding, an idea championed by Colusa newspaper editor, Will Green. He strongly opposed the Corps' narrow channel approach to flood management. His idea eventually gained support in the Progressive era, and the Sacramento River Flood Control Project was enacted by Congress in 1920. Today, when driving over the Yolo Bypass between the cities of Davis and Sacramento, we are looking at a simple system that diverts a

At the time of these immense public water works, little thought was given to negative effects on fish and wildlife.

Looking down Shasta Dam

tremendous amount of flood water away from the city of Sacramento. Several other nearby bypasses protect land and people the same way.

The comprehensive planning period of water development accelerated during the Great Depression. The state proposed a massive and expensive state Central Valley water project and voters in the early 1930s actually approved the idea. However, the state could not sell the bonds during the Depression. So the federal government stepped in, and the huge Central Valley Project (CVP) was created under the new Roosevelt administration. When Shasta Dam was built at the north end of the Sacramento Valley in the 1940s, the water in the Sacramento River was finally managed during times of drought and flood.

The dream of the early settlers to control and use the water of the Sacramento River for farms and cities finally was realized.

At the time of these immense public water works, little thought was given to negative effects on fish and wildlife. By 1980, the salmon and other fish were diminishing mainly due to dams and water diversions. Concern for the fish, along with the new attention to environmental degradation across the country, led to the growth of a new social movement. Those growing environmental issues would be part of the next big water fight in the Sacramento Valley and the rest of California.

But that's another water story.

Sacramento Valley

The Orchard

Stephanie Taylor

In its prime, this 20-acre orchard yielded 40,000 pounds of almonds a year.

An almond orchard lies in destruction. Fragments of branches and tree trunks are piled up like gravestones in a cemetery - a monument to each tree. In the midst of the field stands a farm worker and a backhoe. The sight of the decimated field is disturbing. Or maybe I am ashamed that I understand so little about the cycle of agriculture in the Sacramento Valley.

Past the orchard is a farmhouse and garages full of equipment. Nearby are other orchards, each in different stages of maturity—some young, some old, but all with the bare branches of winter reaching to the promise of warmer skies in spring. Each two-toned trunk is set in neat rows—a familiar sight for those who pass through the Valley. I interrupt the farm worker, a tiny tree in one hand, to learn about the orchard that has been cut down. He calls the owner for me.

Gloria Lopez is a third-generation farmer. She grew up on this land, working beside her family from age 6. She remembers her father planting this orchard in the early 1960s, grafting each almond start to a peach tree root for stability. With admiration and love, she described her father as an innovator, architect, engineer and a farmer who managed his property astutely. And this was the last remaining orchard Sebastian Lopez had planted.

Sebastian Lopez was the son of Spanish immigrants who, along with 8,000 others, left Andalusia as peasants in 1907, having been promised work on Hawaiian pineapple plantations. Promises unfilled, his family left Hawaii for San Francisco, and after the 1906 quake, settled in Winters, drawn to the Mediterranean climate. Working as field laborers, they saved and saved.

The Depression hit, and with their savings, they bought land that farmers had mortgaged and lost to banks. They worked hard, farming apricots, peaches and almonds.

Gloria and her husband, Mike, joined her parents on the farm. In its prime, this 20-acre orchard had yielded 40,000 pounds of almonds a year. As the trees aged, production declined to 6,000 pounds. And four months ago, Sebastian Lopez passed gently at age 89.

The piles of almond trees will be sold as firewood. Thirteen-hundred walnut trees are planned for the now-vacant 20 acres. Gloria believes that, in the Sacramento Valley, walnuts offer a higher yielding, more reliable crop. For now, those stacked monuments stand as a testament to Gloria's father and three generations. The walnut orchard will renew the cycle of agriculture in the Valley, but Gloria wonders if her family farm will survive for another generation.

Four years later, I visited Gloria to see how she and the orchard were faring. She had recently retired after 35 years as a college teacher. She'd written a wonderful book about the Spanish families in the Winters area, *An American Paella.*

We sat at her kitchen table, looking out to her garden and the orchards beyond. Her husband, a large man with the huge, worn hands of a farmer, came in from the orchard. We talked about family farms versus big agri-business, and what it takes to survive today: planning for succession in family businesses, the importance of farmers working together, wells, water, and what they'd learned from Sebastian. The reason, they said, that they've been able to continue farming, is that they never mortgaged the land, the land that he paid for.

We walked out to the flower garden. Gloria said her father was always annoyed with her mother's gardens, asking "Why ya gonna water something you can't eat?"

We strolled out into the orchards. What had been piles of almond refuse was now a thriving walnut orchard, new bare branches reaching to a sky blue day. Rows of young trees, separated by freshly mown grass, stretched west towards the Coast Range. Trunks painted white to protect against sunburn, were the only stark white in the landscape besides giant cumulous clouds bearing a false promise of rain.

Sebastian's words, "Why ya gonna water something you can't eat?" echoed, and morphed into "What ya gonna eat if ya can't water?"

An abandoned orchard

Rice: Turning the Story Around

Rita Schmidt Sudman

The future of farming and the rice industry will depend on continuing to find new ways to keep the Valley in profitable and environmentally friendly farming.

Rice is an ancient crop, but relatively new to the Sacramento Valley. Grown in Asia for more than 5,000 years, it came to the new world in 1685 when a storm-tossed ship from Madagascar reached South Carolina. Rice seeds carried on the ship suited the coastal lowland marshes and soon the profitable crop was called Carolina Gold.

After the California Gold Rush, about 40,000 Chinese, and later Japanese immigrants, created a strong demand for rice. Attempts to grow rice in California failed throughout the 1800s, and had to be imported, which was expensive. Finally in 1908, W.W. Mackie, a government soil specialist, conducted the first successful trials to grow rice in the Valley's Sutter Basin. In spite of the area's poor soils, he theorized that the area would be "the best rice producing land in the world" because of the overflow water from the Sacramento River system. Japanese immigrant Kenju Ikuta then demonstrated that rice could be produced commercially at the Biggs Rice Experiment Station in Butte County. By 1920, California became a major rice growing state and today produces 20% of the U.S. rice crop.

For rice farmers, things progressed for the next few decades. Valley farmers and water districts established senior water rights that gave them dependable and inexpensive water, especially after the building of the federal Central Valley Project (CVP) in the 1940s. At the top of the Valley geographically, the CVP's federal Shasta Dam is the largest reservoir in California.

Then came the environmental movement.

In the late 1970s, the California rice industry came under attack for its farming and environmental practices. The most well-known critic of the industry was author Marc Reisner who came on the scene in the late 1980s and redefined the water debate with his popular book, *Cadillac Desert*. He chronicled the development of water in the West through the glory dam building years and made the case for a reversal in federal water policy.

Flooded rice field in the Sacramento Valley

When I interviewed Reisner for *Western Water* magazine in 1991, he spelled out his arguments.

Reisner targeted four crops which he deemed to be of relatively low financial value in relation to their water use in California: cotton, rice, alfalfa and irrigated pasture. Partly based on the popularity of Reisner's book, the public debate started to center on future solutions coming from increased water conservation, water recycling and water market transfers. Reisner argued that federally price-supported rice was being grown with copious (2 ½ million acre-feet) of low cost water to farmers in the semi-arid Sacramento Valley and was thus damaging the native salmon.

In the 1980s, author Marc Reisner argued that federally price-supported rice was being grown with copious amounts of low cost water in the Valley, and thus damaging the native salmon.

I asked Reisner what the rice growing areas would do without this crop that grows in the Valley's poor clay soils. As background, this was in the recession of the early 1990s when Butte County had declared bankruptcy. He said he favored federal subsidies for new water saving technologies rather than government support for rice and low cost water. And he noted that water standing in rice fields during the Valley's 100-degree summer days causes water evaporation. It was time, he said, for a new way of thinking. "We either have to go out and build mammoth new dams or we have to become more efficient," he said.

His arguments contributed to the changing political climate. In 1992, Congress passed the Central Valley Project Improvement Act (CVPIA) which over time substantially raised the cost of federal water from about $30-100 for water delivered in abundant water years to about $500 an acre-foot for farmers in the drought year of 2014.

Besides water use, rice farmers also were coming under scrutiny for practices causing air pollution. Their longtime tradition of burning the fall harvested rice fields to avoid disease was impacting air quality. I remember moving to Sacramento in the fall of 1978 and seeing the city sky turn dark with air filled with smoke due to extensive burning. After much debate, state legislation was passed to phase out, between 1990 and 2000, rice field burning. Reluctantly, the farmers began incorporating discarded rice straw into soil during the winter flooding thus creating more wetlands. About half of the more than 400,000 acres of rice lands are now also part-time wetlands in a normal water year, according to the California Rice Commission.

Rice farmers made lemonade out of lemons. Their reluctant change in farming practices led to a public relations coup.

There would be only about one-third the current amount of wetlands in the Valley if the burning had continued. Wildlife experts believe this increase in wetlands reversed years of long decline in wintering waterfowl and are a critical element of the Pacific Flyway. Nearly 60% of the winter diet for millions of ducks and geese in the region comes from rice fields. Researchers have identified nearly 230 wildlife species that depend on California rice fields. Environmental groups now applaud the new farming practices. As rice land burning became a farming dinosaur, some farmers turned to running profitable duck hunting clubs that charge $1,500 or more per person per season.

Although rice still consumes about the same amount of water, rice farmers are now seen as environmental stewards.

Clearly, rice farmers made lemonade out of lemons. Their reluctant change in farming practices led to a public relations coup. And although rice still consumes about the same amount of water to grow a crop, rice farmers are now seen as environmental stewards rather than water guzzling and air polluting barons of the land.

When the debate about growing rice in California was raging 25 years ago, no one saw the growth in popularity of Asian cooking and sushi. Nearly all sushi rice consumed in the U.S. comes from California, largely from the Sacramento Valley. California rice is a now a $500 billion industry, second worldwide only to Thailand in exports of premium rice. In fact, the growth of Asian food trends has led to the rice phenomena being called another California Gold Rush, a new Carolina Gold.

Drought also has changed the amount of rice farmed in the Valley. In 2014, rice declined almost 25 percent from the previous year – one billion pounds – due to lack of water availability. In 2015, rice farmers lost 25% of their federal water. In turn, wetlands suffered about the same percentage loss of water.

In recent years, Valley water districts have entered into agreements with farmers to fallow land, and instead, sell the water to San Joaquin Valley farmers and Southern California water users. In 2015, those water districts paid rice farmers $700 an acre-foot – up from $500 in the drought year of 2014, and up from $200 in the normal water year of 2012.

It's more profitable for some rice farmers to sell their water in certain years rather than plant rice. In 2015, about 30% less rice was planted in the Sacramento Valley. In 2016, there was enough water available for farmers to plant a full rice crop.

Another concern for rice farmers is the growing acreage planted in almonds, now California's most valuable crop. The value of the almond harvest in 2014 was about six times the value of the 2012 rice harvest. Unlike rice, almonds are a permanent crop that requires consistent water all year long. Some Sacramento Valley rice farmers are willing to sell water to almond farmers. Although rice

A rice plant

The value of the almond harvest in 2014 was about six times the value of the 2012 rice harvest.

farmers say they don't want to pit one crop against another in the water debate, some farmers and community leaders say there will be increasing negative effects on people and businesses dependent on rice farming - rice mills, dryers and small town businesses—if more water is transferred from the Valley to other parts of California.

The Sacramento Valley - unlike the San Joaquin Valley - has fairly full groundwater basins. Farmers can turn to groundwater if they don't have enough surface supply, or sell surface water. As California's statewide groundwater management law, passed in 2014, begins to take effect, Sacramento Valley farmers will need to better manage groundwater and surface water as one resource. Also, Valley farm interests hope that a new proposed dam, Sites Reservoir, will be built north of Sacramento to provide more flexibility in water management.

These fixes will not be easy. This time around, there won't be huge federal subsidies for the Sacramento Valley as in the past. Rice farmers will have to step up to the table and pay a significant portion of the price for any new water facilities. The future of the Sacramento Valley and the rice industry will depend on continuing to find new ways to keep the Valley in profitable and environmentally friendly farming.

Above: Rice stubble in harvested field. opposite page: Rice storage

Winter Birds

Stephanie Taylor

Honking fills the air to join cries, coos, trills, quacking and the splashing of countless restless wings."

From the dark a startled crane rises, a specter with wings spread wide, glowing in a near full moon rising. He is my reward for a patient afternoon watching and listening to birds.

Millions of birds—more than 200 species traveling the Pacific Flyway, from Mexico to the Arctic Circle—land in the Central Valley during winter. Wetlands restoration has reinvigorated this magnificent phenomenon after much of the birds' habitat was lost to urban development and agricultural practices.

On Staten Island in the Delta, a pair of Sandhill cranes hunts, delicately picking their way through remains of a recent harvest. Other fields are flooded, where thousands of birds float, resting. Pudgy coots feed on the shore, and in periodic panic, flee upon the water with a sound like a small tsunami.

I vow to stay past dark.

Out of the setting sun, vast formations of geese pass. Honking fills the air to join cries, coos, trills, quacking and the splashing of countless restless wings. With night, all that remains discernable is their cacophony.

On the Cosumnes River, at a managed wetlands preserve in Galt, herons roosting stoically in the trees allow our canoe to glide silently beneath.

This wild river is perfectly still. In nearby fields, silence is broken by the calls of Sandhill cranes. I can't see them but their sound is as distinctive as their notorious flapping and leaping mating rituals.

North of Marysville, with the Sutter Buttes as backdrop, Tundra swans cover a farmer's flooded rice fields, white bodies glistening in the sun. When swans sleep, 10 percent stand guard. The sleepy ones coil into clever balls against the chilly morning.

Ever since rice has been grown in the Valley, fields were burned after harvest, destroying bird habitat and filling the air with smoke. Rice farmers have responded to 1991 legislation to improve air quality. Flooding the fields is an effective alternative and the bird populations have returned.

The relationship between birds, farmers and wetlands managers provides an inspiring example that collaborative resource management can benefit polarized parties.

Near Elverta, farmer Jack DeWit and I stroll leisurely along narrow levees. It's early afternoon and nearby hunters have scared most of the fowl into other fields. He's a first generation rice farmer, and like others, he burned his fields after harvest. He's a businessman and keeps an eye on the bottom line. "It costs pennies an acre and the cost of a match," he said. Now, with chopping, discing, flooding, plus the cost of water, fuel and electricity, it's $90 per acre.

He wasn't complaining.

Rice farmers replaced the element of fire with that of water. Water suppresses weeds and diseases that would affect next year's crop. Restored wetlands and left over rice attracts millions of birds. They chew, pulverize and enhance decomposition of stubborn stubble in the heavy clay soil. They eat bugs. They rest, feed, breed, and next season they'll come back.

It's a magical partnership.

Coalescing Around our Values in Water

David J. Guy
President,
Northern California Water Association

Today you will see fishery biologists talking proudly about the preservation of farming and refuges, while the farmers will be talking about the importance of salmon and birds.

The water wars have dominated much of the discourse surrounding California water—yet, to me, the more interesting and salient story is how people come together around water and its special attributes. Yes, water will always be for fighting as Mark Twain quipped, but I admire the thoughtful and innovative water leaders who have crossed various divides to avoid or resolve the wars, typically by finding an alternate route where people can coalesce their thinking around our precious water resources. As the authors of this book, Rita and Stephanie capture this philosophy in different ways through their vivid images and words.

We are truly blessed in California with a singular combination of amazing landscapes, sun and water. In response, people for more than a century have joined together to organize in many ways around water systems--both the water supply and flood protection system--to address the famine and feast that shapes our tenuous relationship with water in this state. Most of my time over the past several decades has been devoted to learning about the Sacramento Valley (the northern part of the Great Central Valley) by listening in the coffee shops and board rooms, walking along the ditches and rivers, fishing its waters and watching the birds fly at sunset. In these settings, I have observed first-hand the way people work together around water.

The first is that people have come together through the advancement of a physical, integrated water management system nested within a truly unique landscape. Every drop of water in the Sacramento Valley follows a flow-path designed to both protect and serve multiple beneficial purposes that reflect our society's current and evolving values in water: cities and rural communities, two million acres of family farms, the national and state wildlife refuges and managed wetlands, and recreation. In the Sacramento Valley, the rich mosaic of these lands are all intermingled, with water flowing through this system in a concerted manner to serve all these beneficial purposes.

A snapshot of the Sacramento Valley today will reveal an increasingly sophisticated managed water system nested within a truly unique landscape. Water is diverted from the rivers through state-of-the-art fish screens, which allow the diversion of water while protecting and keeping migrating salmon and other fish in the river. Additionally, many water supply canals now siphon under creeks and other watercourses to allow fish passage and more effective flood protection. The delivery facilities are operated to greater precision by computers so the water is used efficiently by cities and rural communities, farms, wildlife refuges and other managed wetlands. In the Sacramento Valley, the rich mosaic of these lands are all intermingled, with water flowing through this system in a concerted manner to serve all these beneficial purposes.

Second, while the physical system has been changing, the mindset and approach around water resources has also been evolving in parallel, which has also brought people together. For most of the 20th century, efforts focused on how to divert water onto the land. In the 21st century, the leaders in the Sacramento Valley have recognized there have been consequences resulting from these actions and they are now at the forefront to better understand this dynamic and to carefully integrate the flood and water supply systems and the landscape in a way that works better for both people and the environment in a state with 39 million people. This includes in-stream strategies through agreements, decrees and decisions on every watercourse in the Sacramento Valley that are designed for salmon. It also includes out-of-stream strategies to spread water out over the land (bypasses, farmland and wetlands) in a careful and timely way for farming, to serve birds and the Pacific Flyway, producing food and safe habitat for fish, providing groundwater recharge, and to provide public safety.

Today, when visiting the Sacramento Valley, you will see fishery biologists talking proudly about the efforts to preserve farming and refuges; while the farmers will be talking about the importance of salmon and birds. People have and will continue to evolve in the way they approach water resources as they coalesce around different values that are all important to California. There will continue to be water wars, but let's take time to celebrate the way we come together around water.

Sacramento Valley

Rice Reflections

Stephanie Taylor

North of Sacramento, a tangle of conveyance pipes, tubes, augers and funnels transport rice in ordered chaos.

On winter days, when fog shrouds the Central Valley, immense vertical structures emerge ghost-like from a horizontal world. On hot summer days, shimmering cylinders jut from the flat landscape. They stand like sentinels up and down Highway 99 and Interstate 5, and we barely notice them.

I've always called them silos. A rice farmer scolds me: "Those aren't silos, those are dryers." And when I suggest that as architecture, they're aesthetically impressive, she laughs. To her they're not beautiful, just functional.

Silos, dryers, mills, warehouses are an integral connection between our fields and ports, between California resources and the global economy. They store, dry and process hundreds of products, from rice north of Sacramento to cement for oil wells near Bakersfield. Milo, corn, wheat, barley, alfalfa, sunflower, seed, hay and pond binder, silica, calcium chloride, gypsum, pet food and cat litter.

Understanding their function only enhances my appreciation of their form.

In the rice fields north of Sacramento, Tom Reese climbs into his giant red harvester, starts the engine and heads south across a laser flattened landscape covered in gold and green stalks heavy with grains. At the end of the field, he turns and heads north.

Back and forth.

To the east, the Sierra Nevada provides a backdrop to this contemplative task. To the north, the Sutter Buttes hide behind hedgerows, and to the west, the Coast Range disappears, lost in a waning afternoon.

Conveyers move rice from dryers to storage

We revere the natural landscapes of California, mountains and coast. Too often we take for granted the simple, flat world we see in between. The sublime can be found in our own backyard, in fields and farms and the mechanization of a rice harvest every fall.

Reese, a jovial man with huge hands, has worked this land for 35 years. He describes the process as I lean forward in a seat next to him, and watch rotating combs suck stalks of rice into the harvester. A cylinder separates grains, flinging them into a shaker pan and fans blow the unwanted parts out the back. Another machine keeps pace as rice flows from a chute on the harvester into a cart.

Back and forth.

Across the valley, other rice farmers in other harvesters deliver grains to carts that in turn, transfer rice to waiting trucks. The trucks drive south to Van Dyke's Rice Dryer in Pleasant Grove. Jim Van Dyke's great-grandfather first started farming wheat here in 1898, on hardpan land better suited to rice. His son started the dryer business, and in turn, sold it to his son Jim in 1978. Now Jim's daughters are taking over the immense facility that connects farmers to markets.

About 50 farmers deliver more than 1 million pounds of rice each season to Van Dyke's. A truck unloads green rice into a pit where the grain offloads through grates in the ground and begins an up and down journey to dryers and storage. Rice leaves the fields with a moisture content of 18 to 24 percent and must be dried to exactly 13.5 percent for maximum value and minimum spoilage. Propane fans surround each metal structure and blow warm air into and up through the rice. It smells as if it's beginning to roast.

Harvesting rice

Revolving buckets lift the grains to the top of tall conduits to deliver the rice into huge storage containers, until it's sold.

I can hear the grains flowing, cascading. It's a soft sound, like the pouring of sand.

Back in the rice field, the western sky appears red orange through the dusty sunset. The field sends up an aroma of newly mowed straw. Harvester lights appear across the darkening landscape as farmers continue their work into the night. Reese is ready to go home, ready to leave his field that is quickly becoming too damp to harvest. He pulls to a stop, reaches into an ice chest and offers me a beer.

We stand together watching a brilliant sky fade. I breathe in that smell, the smell of fall in this valley. Earth gives up its heat, creating a fog that hugs the ground. Damp air drifts around us like ethereal ghosts. Blackbirds swarm over cattails that border the fields.

I am awed by this simple experience. Right in my own backyard.

The Coast

Tides

River Beach

Lagoon

Golden Gate

Brackish water

Salt Water

Crab Pots

Fishermen

Oyster Beds

The Coast

The Fisherman

Stephanie Taylor

Dungeness crab is a California tradition, and now that I've seen what it takes to get them to market, I'll never question prices again.

His face was etched with history, leather bound, rough around the edges. Busy on a boat, I interrupted him. He didn't mind, and after my query that I was curious about the crab fishing industry in California, he invited me on board.

I followed him into a cramped, overly warm cabin, a stark contrast to a chilly hint of rain and cloud filled sky above. Getting ready to go fishing, he said. Just inside the door, foul weather gear hung red and yellow, pulled tight with bungee cords. On a narrow galley counter, groceries, cans of soup, bottles of water rested, recently unloaded. I squeezed onto a bench at a tiny galley table.

He introduced himself. He looked older than his forty-six years, the sun, wind, and I guessed more than a few years of very hard living. A full head of dark hair matched a neatly trimmed beard. He wore copper bracelets on both wrists, sleeves of a cobalt blue tee pushed up, and an earring in his left lobe. He wasn't a big man, not as big as I thought a fisherman ought to be, though solid, not beefy, with obvious strength in wrists, arms, neck and back.

He looked a bit like what a pirate might look like, graced with the air of a rebel in blue jeans, a stereotype, a non-conformist, a living-on-the-edges-of-society, itinerant kind of guy.

He confirmed this in degrees, starting with how he began fishing at 21. He was working at a Kmart somewhere close to a harbor as he paused to watch squid boats heading to sea. He had an epiphany, went to the docks in Monterey and signed onto a swordfishing boat. This is how he learned to fish, infatuated with the challenge of landing 400 pound, thrashing swordfish.

This is how he learned to listen, he said, and I thought at the time, what an odd thing for a fisherman to say. Listen to what?

He had blown into Bodega Bay on a violent storm in 1990. 80 miles off Point Arena, it had taken three days to find safe harbor, twelve hours straight at the helm. Exhausted, he staggered to the nearest bar, and has been crewing for local fishing families ever since. Except

for the 10 years he worked as a painter after the trauma of loosing seven friends to the sea—in six years.

We stepped out of the cozy cabin on to the impossibly cramped deck so that he could demonstrate how he catches California's famous Dungeness crab. The season had just started, and he was enthusiastic about prospects for a year of abundance, including high prices. Most years are a struggle, he said, to meet the needs of a precious young daughter, to pay rent, to keep his car working, paying his mechanic off on the honor system. He lived 20 miles away, 10 of them along the steep cliffs of scenic Highway 1.

At the stern, crab traps were piled four and five high, starboard to port. He showed me how he attaches plastic bait traps, identified by uniquely colored buoys. He bent over, and with his knees, lifted a trap to his thighs, and then higher, to show how he hurls them overboard, at 120 pounds empty. I couldn't lift one past my thighs.

He showed me the mechanism that pulls the pots up at 6 feet per second, weighing over 400 pounds full, demanding a man's unfailing focus. He'll catch 7,000 pounds of crab in two days, often working 18 hours straight. He pointed to pole structures, standing like sentries on each side of the mast, that descend at 45 degree angles to stabilize vessels, and said that overloading can roll a boat, pots can pull men to deep deaths. 98% of the accidents are caused by human error, he said.

He was a cowboy in Texas and a sheet metal worker in San Francisco, but he loved fishing and the sea. "It's peaceful," he said. "I wake up every morning at 4:30, put my boots on and look forward to an adventure." He sighed. "I leave my problems on the dock."

The Coast

Salmon and Fishermen: California's Endangered Species

Tim Sloane
Executive Director, Pacific Coast Federation of Fishermen's Associations and Institute for Fisheries Resources

Water diversions deprive fish of the equivalent of the air they breathe. And the folks that benefit from moving all that water, especially the industrial farming operations, aren't likely to give it up any time soon.

I was raised in San Diego, and as a small Californian, only once thought twice about where my water came from. In elementary school the local water district invited my class to submit illustrations for their annual "Water Awareness" calendar. My drawing featured a striated California, deep green in the Northern third, a lighter green in the middle, and an arid tan at the bottom. Little cartoon Reclamation engineers in the North amiably offered water to the desert South. So close, and yet so naively far from reality.

I only started to comprehend the magnitude of California's water issues during law school, when I started working with the Pacific Coast Federation of Fishermen's Associations (PPFFA). That's tragic. It's a failure that any Californian is ignorant of how and why our water system is the way that it is. It's amusing to me, and a little sad, that I never questioned the miracle of growing up so comfortably in a desert. Every Southern Californian should get a water primer in high school.

At PCFFA, I had the opportunity to intern for the inimitable Zeke Grader, the champion of the commercial fishing fleet. Zeke gained renown for forging an alliance between environmentalists and fishermen on California water issues. He was a tenacious advocate for the fleet and the fish, who lived by the belief that both of his constituents had a right to thrive.

Chinook Salmon is the most important species to California's fishing fleet north of Santa Barbara. Chinook used to thrive on the Sacramento, San Joaquin and Klamath Rivers, their tributaries, and most of the coastal streams on California's edge. These fish begin their lives deep in the freshwater system, migrate to the ocean, and then return to their natal waterways to spawn and begin the cycle anew. Fishermen depend on ample, clean, cool water in salmon streams to bolster baby fish populations so they can harvest adults in the ocean. Without healthy juveniles, there aren't many adults on which a fisherman can make a living.

The state's massive engineering project that enables delivery of Northern California water to Southern California's irrigators and cities is the main impediment to strong salmon runs. Dams block salmon migration and prevent passage to over 70% of Chinook's historical range in California. Water diversions that satisfy southern demand deprive fish of the equivalent of the air they breathe. And the folks that benefit from moving all that water, especially the industrial farming operations, aren't likely to give it up any time soon.

The result: our salmon runs are going extinct. And as the salmon go, so goes the salmon fleet.

Salmon have been doing their thing in California for millennia. We've been deluding ourselves that our plan is better for only about 150 years.

I've been honored with the task of succeeding Zeke at PCFFA. It's my duty, and the duty of the fishing and environmental allies I'm lucky to have on my side, to see that fish have a chance in California's fresh waterways so salmon fishermen provide food for California's tables. Even Zeke would tell you that sometimes it's a lot like a fish banging its head against a dam.

But if anything motivates me, it's that nature has a plan. Salmon have been doing their thing in California for millennia. We've been deluding ourselves that our plan is better for only about 150 years. This Southern Californian got the picture. So did fishermen. Here's hoping the rest of California follows suit.

Salmon Troller

Photo: Tim Sloane

The Coast

The Oyster Farm

Stephanie Taylor

Like the California Delta, the silence and beauty of this place belies contentious and complicated issues and emotions.

Winter 1974

The Point Reyes National Seashore has always been one of those California places, so close to the heart, so quiet, so calm. My mother and I visited Johnson's cannery, on the fresh water end of Drakes Bay. Oysters don't come any fresher. She held a shell to her lips, slipped one down, and offered me a precious taste of pungent sea.

I tried, I really did, but it took me several more years and a number of visits back to this wetlands estuary before I acquired an appreciation for such a raw and slimy substance.

Spring 2013

It's early morning at Drake's Bay Oyster Farm and the tide is out. Oystermen push an old wooden boat from a rickety dock and leap aboard. Their leathered faces expressionless, lost in thought, hoods pulled tight, settled down into the chill and the hum of a small outboard engine.

Their day will be long, their labor tedious, their futures uncertain. We head toward the sea, to the oyster beds. From far away, they appear as long dark lines in a landscape of greys, hovering, like a mirage on the water.

Approaching the first bed, clusters of mature oysters dangle from a grid of sagging boards. Other beds lie scattered on the shallow bottom in mesh sacks. There are 19 million oysters out here. More than 50,000 people visit the oyster farm every year to feast and to learn about sustainability, biology and marine agriculture. 500,000 oysters are consumed, rich protein grown with no fresh water. The oysters here generate both tourists and connoisseurs.

Drakes Bay

In a 1961 economic report, the National Park Service acknowledged the “public value” of the oyster farm that “presents exceptional educational opportunities.” Times change; the environmental movement has grown. In 1964, the Wilderness Act passed, and the Park, which owns the land under the cannery, said that the estuary should be protected by the Act, as “an area untrammeled where man himself is a visitor who does not remain.”

Oysters grew in Drakes Estero naturally. They became a high demand, high protein value food source during the Gold Rush. Commercial farming had been formally permitted by the federal government for almost 80 years. When Johnson sold his five acres to the Federal government in 1972, he signed a 40-year lease to farm.

When Drake's Bay Oysters acquired Johnson's business in 2005, they knew that the lease was due to expire in 2012. They asked the federal government to extend the lease, to continue harvesting shellfish. They argued that the estuary should remain "potential wilderness," and that they be allowed to operate, pending the outcome of litigation.

Fractured by the San Andreas Fault, Point Reyes is a peninsula ripped from the California coastline. Moss hangs from trees in dense woodlands, and high on the crest, wind-blown ranch lands offer breathtaking views of the Pacific Ocean. To the south, Drakes Bay remains pretty much as it must have looked in 1579 when Sir Francis Drake arrived. Some believe he repaired his ship in its calm estuary. In its tidal flows, oysters thrive and are prized by regional restaurants.

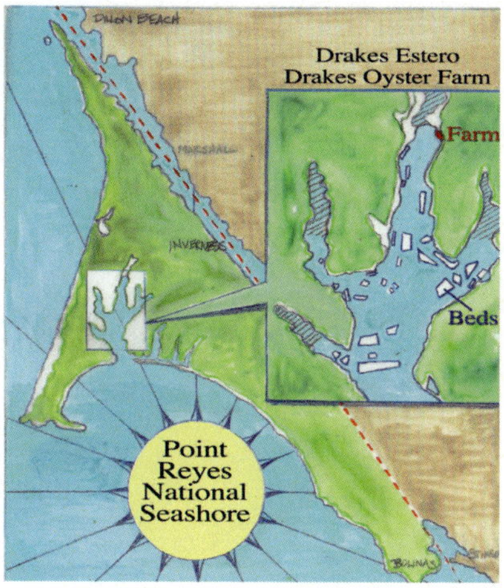

Tying the boat to the oyster racks, the men step cautiously across the narrow wooden boards. One man lifts a batch of oysters strung together on a wire like a necklace. He passes it to another man, who passes it to another, until a small barge is piled high. About an hour later we're back on shore at the cannery, where they'll spend the rest of their day sorting and packing.

It's a timeless process, this plankton to protein, this touch of human hands from seeding to sorting. As filter feeders, the oysters help keep water clean while a diverse ecosystem has responded, surrounding the beds like a reef.

Winter 2015

The court decided not to extend the lease. The farm closed at the end of 2014, and the Park Service began removing what remained of the cannery.

On January 3rd, a few hundred Point Reyes residents gathered for a potluck. Huge barbeques were covered with the very last of the freshly harvested oysters. Family and friends celebrated the history of the oysters, and at the same time, looked forward to a different kind of future.

The Coast

High Tide at Carmel River Beach

Stephanie Taylor

Time: 11:50 a.m.

Between river and sea, wind pushes sand into a dune that separates the river from the sea for most of each year. The river estuary becomes a lagoon, a sanctuary for creatures hiding in reeds. In drought years, water levels drop, the dune grows higher. Hundreds of sea birds gather to bathe, rising, diving, flapping, flying. It's a peaceful place, but not today.

On the ocean side of the dune, eight-foot surf crashes against a steeply eroding beach, sending spray exploding. Wild storms from Hawaii have brought torrents to the lagoon, and carved a mini canyon through the sand. The Carmel River has finally broken through to the sea. Pushed by powerful currents, a plastic bottle is caught for a moment in waves. Pulled, driftwood flows in from a sea aspiring to enter.

The tide is coming in, and will reach mean high at 1:05 p.m.

Fresh water meets salt water.

The dune used to span to the south, providing lovely walks to Highway 1 and Point Lobos. Not now. The river denies access with a ten-foot deep channel, and on the surface, water roils, boiling. Defined by the shape of the beach on each side, waves collide diagonally across a 20-foot span. River rushes out as waves rush in, surging over, sucking under, doing battle.

On the lagoon side of the dune, sound of crashing waves is muted. I can almost hear joy in the river, now that it's reached the sea.

In warm winter sun, birds arrive. About two dozen gulls and one pelican flap their wings against the surface. A massive roller pulses far across the quiet lagoon surface. As it reaches the birds, they rise in unison, in protest, responding to the call of just one gull. The pelican has disappeared.

Time: 12:30 p.m.

Waves are higher, with tide, with wind, an off shore breeze promising power. Last night, between drenching downpours, the lagoon was close to flooding, homes marked with nervous sandbags, and the cacophony of a million tiny tree frogs in wet night air.

Time: 12:45 p.m.

The river has risen a good two feet. Surges bring waves far into the lagoon to lap at crumbling shores. Relentless the waves are, one on top of another, closer, closer. Waves invade, retreat, setting white foam gently upon the sand, a billion bubbles quiver. Quiver with the wind.

Wait for high tide.

Time: 1:05 p.m.

The Coast

Golden Gate: Bridging Water—Salt and Fresh

Stephanie Taylor

The San Andreas Fault lurks near this bridge, splitting the west side from the east. The Pacific Plate to the west creeps north at about the same rate as the growth of your fingernails. To the east, the North American Plate moves south. All four of my grandparents lived in the Bay Area during the 1906 quake. I grew up crossing the Golden Gate, though my only memory of it as a small child is of the toll operator who gave me a Hershey's chocolate bar. I've been waiting for another ever since.

Crossing the Golden Gate Bridge is dangerous. Views distract drivers. Oncoming traffic is too close and it's risky to bike. The best way to understand this glorious structure is to walk. Walking is the only way to fully appreciate its attributes, its relationship to land, city, weather and the water below.

From the south shore, the entire structure looks impossibly long, sensuously curvaceous and very high. Of course we know how high, notoriously high. I began my walk at the south entrance with a keen sense of anticipation. This is the first time I've ever walked the Bridge.

Mapmaker of the American West, John C. Fremont, named the 6,700 foot wide strait in 1846. It reminded him of a harbor in Istanbul called "Chrysopylae," (Golden Horn). Winds often reach 60 mph. Designed to withstand 100 mph, it's only been shut down three times. It can sway up to 27 feet. The architect chose the color International Orange #C0362C for visibility in fog, contrast to the sky, and integration into the land, especially in the setting sun.

As I start north, I look through chain link to steeply sloping land below. An immense concrete abutment disappears far into bedrock and contains enough concrete to pour a five-foot wide sidewalk from San Francisco to New York. The total length of the bridge is 8,981 feet, or 1.7 miles. Walking should take me less than an hour.

The day is remarkably warm, clear and still for February. By the time I get to the other side, 3 ½ enthralling hours will have passed. I'll have met a biologist who tells me about the recent return of Harbor Porpoises, I'll have twisted my body into impossible positions to take pictures. I'll be hungry and in need of a glass of wine on the porch at Cavallo Lodge to meet my ride.

Chain link ends as the land below disappears into water. I'm surprised. Shocked. It's all so open. Open to the vista to the west, from Tiberon to Angel Island, Alcatraz, the Bay Bridge, the City, and most of all, open to the surface of the water far below. The railing needs fresh paint. The color shifts from the deep orange to black. Graffiti marks the surface in spots with the human need to declare, "I was here."

The noise from closely passing vehicles is deafening, especially as tires hit expansion joints in the roadway, over and over. Trucks and motorcycles scream past, but my attention is drawn up to the towers and cables, and soon sound diminishes.

Vertical cables tension between the main cable and the roadway. I can touch them, feel their solidity, see the layers and drips of 75 years of paint. I like the view between them, framing each attraction as I pass by. The famous suspension cables soar upwards to the towers from the railing. 80,000 miles of wire are contained in the main cables, wire woven from 27,572 threads– enough steel to circle the equator three times.

It's on the 820 foot Art Deco towers that the 1,200,000 rivets are most apparent, especially in the setting sun. Each rivet casts its shadow across intense vermillion orange. I've heard that the crew of painters starts at one end, paints to the other side and begins again. This is not what I see as I look up. Paint peels, lifts and cracks off the surface, exposing raw steel to the elements.

A man looks down, searching the surface of the water. He's a biologist tracking just one of more than 130 species of fish in this estuary, the largest on the Pacific West Coast. In addition, 500 species of wildlife join millions of birds in this fertile and vital Bay-Delta. It contains 90 percent of California's wetlands, and functions like the natal unit of a hospital, protecting and nurturing new generations of creatures. He tells me that Harbor Porpoises have returned just in the last few years, after a sixty-five year absence.

By ones and twos they swim to the sea from the Bay as the tide goes out. We see mothers with their babies. They emerge from the depths and disappear again. The scientists know why they left, during WWII with the installation of submarine nets, but they don't know why they've returned now, six generations later.

Building a bridge is a war with the forces of Nature.

- Joseph P. Strauss,
Chief Engineer, Golden Gate Bridge and Highway District

This bridge spans an inflow, outflow of tidal energy. Passing beneath, salt water can reach as far inland as Sacramento, especially in times of drought. Fresh water, including Sierra Nevada and Coast Range watersheds, drains 40 percent of California's landmass, and pushes back. That shifting point of battle is called the salt line. While there's no such thing as "normal" in California, 13 million acre-feet of fresh water can flow through the Delta, under this bridge, and out to sea. And twice that much in years of heavy snow and rain.

Tidal currents exceed 8 feet per second. This is hard to comprehend, but fast enough to sculpt one of the deepest natural channels in the world, at about 375 feet.

The floor of the Bay is constantly changing due to these powerful tidal currents. Far below on the ocean floor just outside the bridge, currents have created giant sand-waves. These dune-like structures are 700 feet wide and 30 feet tall.

I reach the north end and turn to look south, back across the bridge and out into the fading light. The Bay today barely resembles that of the pre-1850s environment. Shallow shores have been expanded 40 percent with landfill, with a loss of 80 percent of the original wetlands. Climate scientists make interactive maps that document what might happen with rising seas.

Like the porpoises, perhaps the water has only been waiting to return.

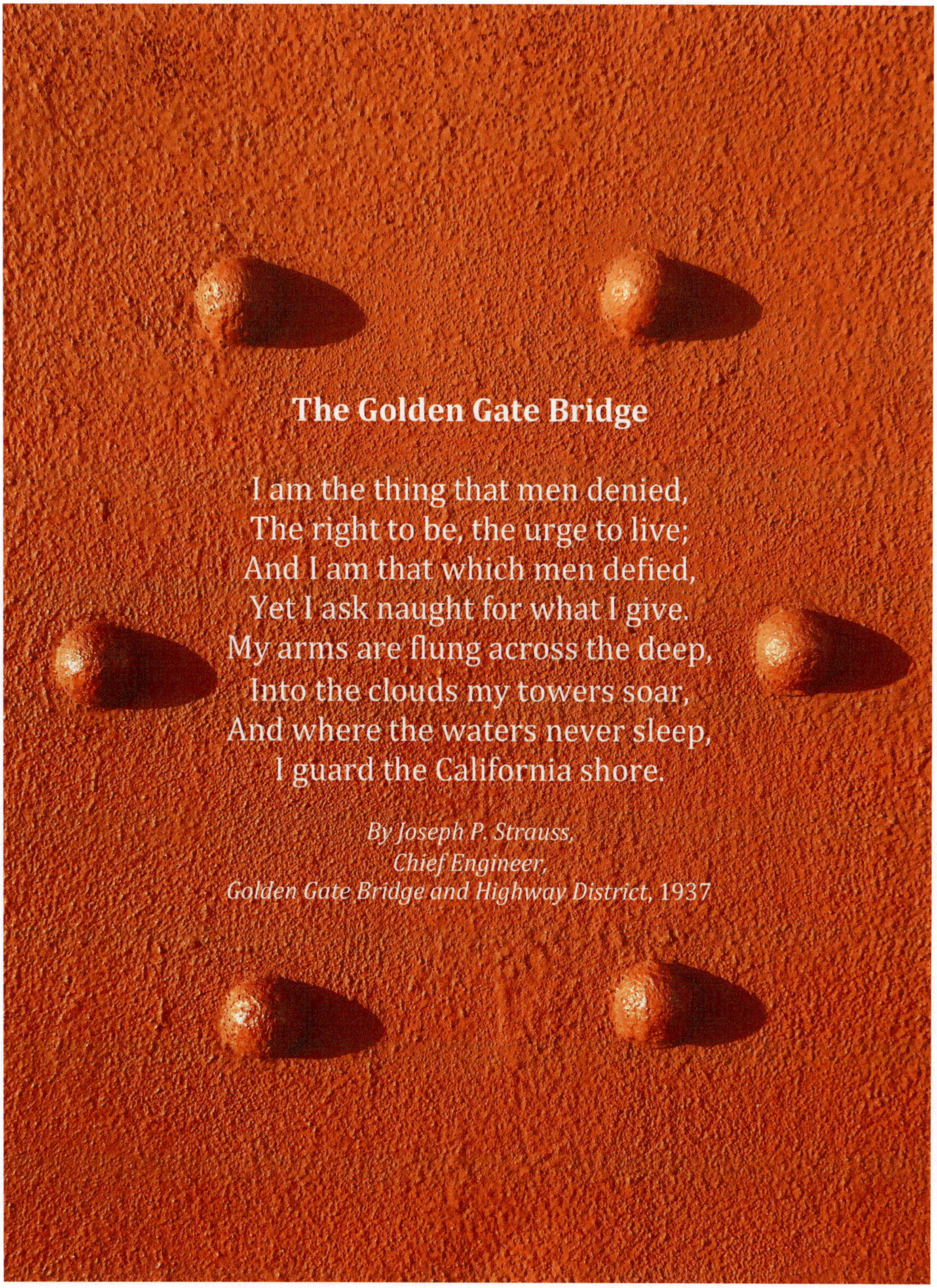

The Golden Gate Bridge

I am the thing that men denied,
The right to be, the urge to live;
And I am that which men defied,
Yet I ask naught for what I give.
My arms are flung across the deep,
Into the clouds my towers soar,
And where the waters never sleep,
I guard the California shore.

By Joseph P. Strauss,
Chief Engineer,
Golden Gate Bridge and Highway District, 1937

The Delta

Estuary

Water Rights

Levees

Habitat

Climate Change

Peripheral Canal

Islands

Trust

Sustainability

Salmon

Tides

Tunnels

Delta Smelt

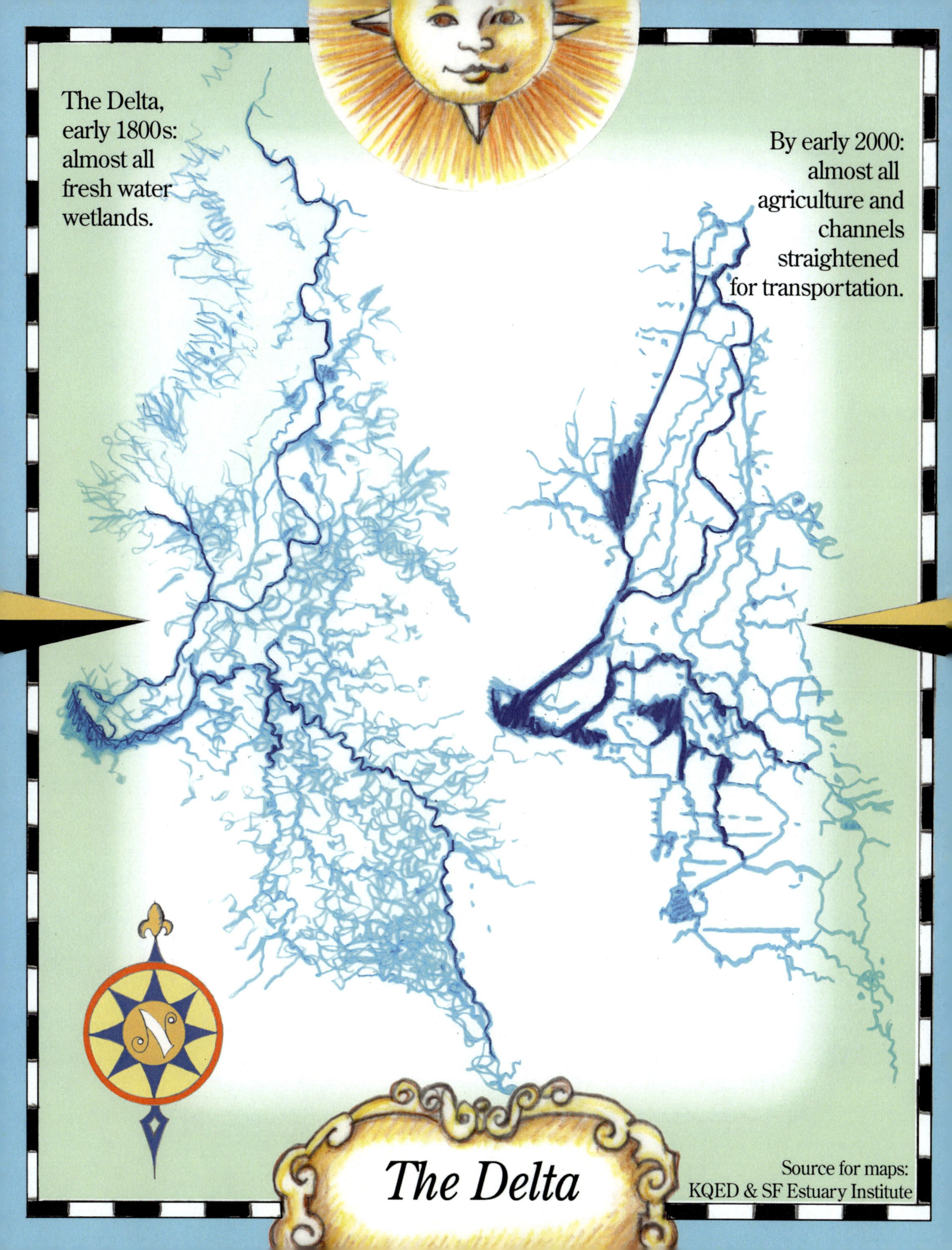

The Delta, early 1800s: almost all fresh water wetlands.
By early 2000: almost all agriculture and channels straightened for transportation.
The Delta
Source for maps: KQED & SF Estuary Institute

The Delta

Lost in the Delta

Stephanie Taylor

> *Fresh water exhaling from our rivers to the Delta is paramount to a healthy system.*

The air is hot, still, silent. The water's surface flawlessly mirrors sky and tules, a dual image of calm that conceals a world of conflict. From the west, tides flow in and out, pushing and pulling. Salt water versus fresh, exotic species versus native.

The Sacramento River watershed in the northern Central Valley and the San Joaquin in the south merge to create a labyrinth of waterways. A vast tangle of sloughs and canals embodies the confusion as to how water from California's most critical watershed, the Sacramento-San Joaquin Delta, should be used and by whom.

From the water's surface, monotonous and fragile levees hide sinking farmland and all navigational landmarks. With 1,100 miles of waterways that would stretch to Kansas, it's easy to get lost.

The Gold Rush irrevocably changed this estuary. Before man harnessed the value of the Delta, it was tidal and seasonal wetland. An old map reveals what resembles a lung-heart system, and that is how I began to think about the Delta. A shifting zone between ocean and rivers, breathing in and out on every tide, twice a day, it is an incubator, a safe haven for plants and animals. With ominous irony, that breath of tide brings salt from the ocean, ruinous to crops and drinking water. Fresh water exhaling from our rivers is paramount to a healthy system.

Now, forever altered, the Delta functions as a gigantic, circulatory system with ever smaller arteries that also sustain the Central Valley and cities farther south. Water flows into or out of the estuary from and to thousands of manmade dams, pumps, canals, and treatment, reservoir and management facilities all over California.

Politics and law have created a shifting tidal zone between policy and science. Delta water is divided between the ocean, agriculture, urban and industry, and those in Southern California. It's more valuable than gold.

Delta Tules

The people who live in this world of water, near it and on it, speak about a sense of community, a lifestyle. The bountiful market in Courtland is a vital symbol where neighbors meet and bond, old families with new. Farmers and fishermen, boaters who dock and those who anchor, all share a passion. It is a place of intensity, from harvest colors to light sparkling on water, to opinions as to where this precious resource should be diverted.

Watch the sun set and the moon rise as darkness shrouds both silence and chaos.

The Central Valley Project's pumps and fish screens near Tracy is my southern most stop on the Delta. Here, water quality is constantly monitored and flows adjusted for levels of fresh water. Conscientious humans rescue smelt and the salmon that have lost their way. Inside, massive pumps lift water to the Delta-Mendota Canal, and gravity pulls it south.

To comprehend this deceptively serene place, walk into a field, smell the earth. Go out on the water, jump in. Listen to the birds and the breezes. Watch the sun set and the moon rise as darkness shrouds both silence and chaos.

Like that period of time between the ebb and flow of tides, Delta survival pauses. As an indicator of the balance for all species in the area, including humans, and amid the chaos of confusing issues and factors, we must avoid getting lost.

Delta sunset looking west to Staten Island and Mt. Diablo

Chasing an Elusive Fix

Rita Schmidt Sudman

The Delta is a microcosm – a world in miniature – of all California water issues.

Spend any time on water issues and you will know that the California Delta is the heart of the state's water system. In my water career of the last three decades, I have seen the emergence of two conflicting philosophies: one of protecting nature and one of subordinating it.

What is the Delta?

The Delta is an estuary where water from mountains and rivers mixes with ocean tidal water causing extreme fluctuations of fresh and salt water. Officially called the Sacramento – San Joaquin River Delta, it is the largest estuary on the West Coast and covers parts of five Counties. It's no longer a natural system, but a 700-mile maze of sloughs and islands protected by over a thousand miles of levees. Within the levees, there are about 60 below-sea-level farm islands.

At the south end of the Delta, state and federal water projects extract water to serve 25 million people in the Bay Area, the San Joaquin Valley, parts of the Central Coast and urban Southern California. It supplies water to more than 3 million acres of farmland, mostly in the San Joaquin Valley.

The Delta's aging levee system can be affected by tides, earthquakes, floods and sea level rise. Environmental problems include invasive species, export pumps pulling fish in the wrong direction, and degraded water quality. The Public Policy Institute of California (PPIC) says that within 50 years, it is highly likely that numerous Delta Islands will fail and flood permanently, pulling tidal salt water into the Delta, adversely affecting farming and jeopardizing much of the state's drinking water.

The Delta is a land in crisis because we humans have changed it.

This situation didn't develop overnight. The Spanish "discovered" the Delta in 1772, and explorers wrote about its abundance of fish, game and fowl. Trappers, including American Jedediah Smith, hunted plentiful beaver.

Photo: CA Department of Water Resources

After California became a state in 1850, the Delta landscape began to change.

The 1855 Swamp and Overflow Act allowed farmers to purchase marshland for $1 an acre to "reclaim" it into farmland. To build the levees to hold back the rivers and floods, farmers used Chinese laborers. Once the mechanical dredge and gas pumps came on the scene, levee building began in earnest. By 1880, there were 100,000 reclaimed acres and by 1930, the entire Delta's 450,000 acres were farmland.

The Delta is battleground zero for the opposing philosophies of protecting nature or subordinating it.

Today the Delta area is home to half a million people living in cities and towns. Its landscape includes highways, natural gas lines, high voltage transmission lines, railroads and two deep water shipping channels.

While the Delta was growing, so was the state, which meant moving water to areas that didn't have enough water.

Soon the Delta would become part of the state's water conveyance system. By 1940, pumping from the Delta began with completion of the federal Contra Costa Canal, the great Central Valley Project's (CVP) first unit. Pumping increased in 1951, when the CVP built its export pumping plant at the south end of the Delta to carry water south to the Central Valley. In 1960, California voters passed bonds for the State Water Project (SWP) and legally committed to exporting more water to Central and Southern California. Within a few years, water was being sent south from the nearby state pumping plant into the new California Aqueduct.

Peripheral Canal proposal

While moving water through the federal and state water projects, planners always considered the Delta as the system's "weak link." There was no Delta facility to efficiently move water through the maze of sloughs. By 1973, a state advisory committee concluded that a joint federal-state peripheral canal, built around the eastern edge of the Delta, was the fix. The canal would isolate the fresh water from salt water intrusion, and protect fish pulled in the wrong direction, toward the pumps, rather than naturally moving west toward the ocean. By 1977, the state Department of Water Resources agreed.

Californians got their chance in 1982 to weigh in on the proposed Delta project. I covered the contentious vote on a referendum that defeated the Peripheral Canal package. For the first time in California history, a statewide water proposal failed and the defeat shocked the water development community.

Many water exporters are devising ways to reduce their reliance on the Delta. This is a wise move.

During these years the environmental movement grew and so did public interest in the Delta.

Since the 1980s, the environment of the Delta has continued to deteriorate. Delta fish species have been listed as threatened and endangered under federal and state laws. Pumping from the Delta was at times limited, sometimes severely affecting the junior water rights farm interests on the west side of the San Joaquin Valley.

About 25% of all anadromous fish, including four runs of salmon, journey through the Delta to spawn. Eighty percent of the state's commercial fishery species live in or travel through the area. The popular striped bass, introduced from the East in 1879, is the centerpiece of many local fishing tournaments.

Why should we care about smelt?

One fish that has received lots of attention is the Delta smelt because it is almost extinct. The fish once was so abundant it was harvested commercially. The smelt is a federal and state protected fish that has declined because of a number of factors, including its proximity to the pumps. A 2015 index survey by state scientists measured the abundance of the fish at zero. The three-inch fish is found only in the Delta, lives one year and likes to swim near the export pumps. Why care about this little fish? Scientists say the Delta smelt is like the "canary in the coal mine," is an indicator species, and a warning of the unsustainable way water is pumped out of the Delta.

There have been many political processes focused on solving the Delta conveyance and habitat problems as the public became more interested in environmental issues. In 1992, one of the goals of the CVP was changed by Congress to include allocating additional water for the environment, the first time a major project had to consider the environment equal to its other purposes.

Strong legal assurances of Delta protections – almost written in blood - seem necessary if any conveyance will ever be built.

Diverse interests seek acceptable scientific and political Delta solutions

In my years with the Water Education Foundation, I covered many processes bringing diverse interests together to find acceptable scientific and political Delta solutions. There have been an abundance of Delta processes, commissions, blue ribbon task forces, strategic plans, scientific studies, and appointed panels!

I've interviewed and discussed water issues with five governors and watched each struggle with the Delta problem.

1 During his first governorship, Democrat Jerry Brown supported the Peripheral Canal proposal, though showing little enthusiasm.

2 Republican George Deukmejian supported a natural Delta conveyance system, called Duke's Ditch, which never got off the ground.

3 Republican Pete Wilson had more success working with the Clinton Administration to create a Delta negotiation process and the joint state-federal CALFED program. It promoted habitat restoration and an integrated view of water issues before it ran out of steam.

4 Democrat Gray Davis continued to try, mostly unsuccessfully, to keep the CALFED program afloat, despite a change of administration that put a Republican president in the White House.

5 Republican Arnold Schwarzenegger created a Delta Vision Blue Ribbon Task Force whose members wrote a strategic plan and began a huge habitat plan, the Bay Delta Conservation Plan (BDCP).

6 In his latest term as governor, Jerry Brown, has revamped the BDCP and has strongly supported a project to construct two large tunnels under the Delta to move water to the state's California Aqueduct with less harm to fish.

Of all the processes, the CALFED program was a game changer.

The idea was to address the Delta system as a whole watershed, ranging from Redding to San Diego. For the first time, scientists were brought into the discussions. CALFED funded a number of successful projects that increased water supply through conservation, recycling and groundwater storage. About 130,000 acres of habitat were acquired.

The medicine was good, but the patient continued to decline.

Co-equal goals for the Delta

Along the way, the stakeholders' pursuit of a solution led to agreement on the co-equal goals of water reliability and ecosystem improvement for sustainable Delta management. In 2009, the state Legislature passed a package of water bills and created the Delta Stewardship Council, charged with designing ways the co-equal goals would be achieved.

The Bay Delta Conservation Plan (BDCP), 2006, was to create restoration conditions that would lead to long term (50 years) federal permits to export water from the Delta. The 50-year permit made the program extremely attractive to water contractors because of the reliability it promised.

As part of the deal, the plan included almost 100,000 square miles of habitat restoration. A similar approach had worked on the Colorado River where a habitat plan was developed for a 75-year permit that allowed water to continue flowing to Southern California.

In 2015, Governor Brown announced a revamp of the BDCP program, cutting the restoration down to 30,000 acres. When he announced the change, Brown stressed the importance of this program to improving the state's infrastructure saying, "The very fabric of modern California is at stake." While this plan, now called California EcoRestore, is more achievable, it will not lead to a 50-year permit. The tunnels project remains separate.

A 2004 levee break in mid-summer flooded the below sea level island of Jones Tract, destroying farms and infrastructure.

Photo: CA Department of Water Resources

Reducing reliance on the Delta and opposition to Tunnels proposal

While key water exporters remain committed to the new process, many also are devising ways to reduce their reliance on the Delta. Maybe this is a wise move. However, it's unlikely that Southern California and Central California can ever completely eliminate their need for water that passes through the Delta. So, some type of acceptable conveyance is necessary.

When an emergency happens and levees fail, actions will be taken without a well thought out plan.

Attempts to provide a way to convey water through the Delta have been opposed by local Delta interests. Many Delta farmers have senior water rights and essentially take water for free. Their concern is that the tunnels will separate water for export from the common pool of water and remove state and federal incentives to release water from upstream dams to keep all Delta water fresh and usable for their crops.

In late 2016, long time Delta observers from the Public Policy Institute of California's Water Policy Center suggested a "grand compromise" to reduce conflict, litigation and possibly resolve Delta water problems. They suggest building one tunnel, not two, to reduce and cap the amount of water taken from the Sacramento River. To benefit Delta residents and make them partners in the water delivery system, they would provide local access to tunnel water in places were Delta supplies are salty and they want the state to continue to help locals strengthen levees. To benefit the enviroment, they want to see a flexibly managed, guaranteed block of environmental water to create a more functional and sustainable estuary.

Trust between stakeholders is essential.

Perhaps this or other compromises would lead to the one element missing - trust between the negotiating stakeholders - people with fundamentally different philosophies. Unless there is some trust among these stakeholders, there cannot be a deal or agreement on any Delta solution. Strong legal assurances of Delta protections—almost written in blood—seem necessary if any conveyance will ever be built. For Delta interests, it is a question of who will govern any conveyance project. And until one or the other side of these prevailing interests wins the public debate, Delta solutions for habitat and water conveyance remain elusive.

Around the period of the First World War, a political philosophy rose to oppose monopolies like the railroads, and instead use government to build water projects that were failing under local control. Today we face two opposing political philosophies on how to move water through the Delta in an environmentally friendly way. It's probable that the status quo will not hold, and the physical Delta will fail within our lifetimes or our children's. It will fail because of sea level rise, earthquakes and environmental degradation. Levee failure will affect infrastructure and cost billions to rebuild while jeopardizing the state economy and environment. When the emergency happens, actions will be taken without a well-thought-out plan.

Everyone will lose.

Photo: CA Department of Water Resources

Some Days are Better Than Others

Campbell Ingram
Executive Officer, Delta Conservancy

It has taken me years to overcome my naiveté of expecting rapid comprehensive solutions to such complex problems.

Some days are better than others. One day you feel like you are making real progress on challenges that are critical to California's survival, and the next you can feel like the problems are unsolvable and that nothing will change unless something catastrophic happens - say the drought persists for another four or more years, or we lose the west Delta and it becomes a saline inland sea.

I'm not alone but I am one of the relatively few who have worked directly on all of the recent efforts to deal with California's water issues. From implementing CVPIA programs, to working on the CALFED EIS/EIR and then the Ecosystem Restoration Program and Environmental Water Program, to negotiating the Bay Delta Conservation Plan's restoration objectives, to now being an implementer of the Delta Plan and California Eco Restore with Proposition 1 funding.

That's a lot of water under the bridge, pun intended, and a lot of my life spent on programs that arguably have not been terribly effective at improving our water supply or the ecosystem on which we all depend. I've seen the worst of our natural tendency to protect our interests and the best of collaboration - but still no durable solutions.

So I ask myself, often, why keep at it, why not go work on something with more tangible successes? It's a hard question for sure but it comes down to a couple of factors.

First, I think most of us who do this work are optimists at heart, and we believe there is a path forward and are willing to work hard to find it. Next, there is something enticing about working on something so fundamentally important to our way of life. There has been wonderful energy in many of the programs listed above, an electricity that comes with thinking out of the box and trying to solve challenging problems. I suspect that after 16 years of working on these issues, anything less critical would not hold my interest.

A levee protects a below sea level Delta island

And there is slow steady progress - while the above-mentioned programs have not achieved all of their objectives, they have all had successes and each has made some progress toward sustainable management. It has taken me years to overcome my naiveté of expecting rapid comprehensive solutions to such complex problems. This is probably one of the biggest challenges to me personally, balancing the understanding that slow progress is reality, with the urgency of failing ecosystems and water supply.

Bringing all interests together to engage information in much more robust ways in real time, all looking at the same data sets and working together to make decisions: that is transformative.

Moving forward, I see two bright spots as we continue on our water policy journey; we are tending more toward true collaboration with all interests invited to the discussion; and we are starting to take advantage of technology that allows us to get more information out of the myriad data sets and modeling output we have invested in over the years. Bringing all interests together to engage information in much more robust ways in real time, all looking at the same data sets and working together to make decisions: that is transformative.

A Glass Half Full or Half Empty Depends on Action and Leadership

Sunne Wright McPeak
President, Delta Vision Foundation, and CEO, California Emerging Technology Fund

Hydrographs for the last century show that only about 3 years out of every 20 are 'average.

The title of this book "Water: More or Less" is reminiscent of the philosophical question, "Is the glass half-full or half-empty?" Each of us has a view on that question and a perspective about water in general that is filtered through the lens of our experiences. I grew up on a dairy farm in the San Joaquin Valley where my father planted alfalfa for the cows, often being "given" the water by the irrigation district scheduled to irrigate in the middle of the night, so I know what it is like when the Ditch Tender and Ag Inspector are the most powerful people in your life.

I've lived in the Bay Area for 45 years, fished the Delta, and was an elected County Supervisor in the region, so I have a deep appreciation for the sacred natural beauty of the Bay-Delta Estuary. And I served as Secretary of the State of California Business, Transportation and Housing Agency, so I know the imperative for all regions to have sufficient, clean water to support the state's economy and quality of life for all residents.

As Mark Twain is famously reputed to have quipped "whiskey is for drinking, and water is for fighting," California's history is replete with various episodes of the "water wars." I, too, have been a part of a few major battles in the last four decades. In pondering the question "More or Less?" it is instructive to realize that the very first session of the California Legislature, after being admitted to the Union in 1850, was dominated by a debate over water rights—farmers versus miners, over heavy metals from gold mining washing down the rivers and contaminating water for irrigation.

Although the discovery of gold was a leading factor in California being granted statehood (by a margin of only one vote in Congress)—the farmers won that battle in the Legislature—forever defining agriculture as bedrock for California's economy and embedding water issues into California's political DNA. Yet, in spite of a succession of water wars, an examination of a map of the California water system is a graphic illustration of how inter-connected we are as a state. Except for the far northeast around the headwaters of the Sacramento River, every region of the state is reliant to some extent on another. That interdependence for survival is why water is so fundamental to California's greatness.

Our interdependence for survival is why water is so fundamental to California's greatness.

So, why has there been so much controversy and dissention about water over the years?

The heart of the problem is the lack of sufficiently insightful and insufficiently sustained leadership to address shorter-term regional interests with a longer-term integrated solution for all regions.

Water supplies for everyone and everything—families, fish, farms and factories—are unreliable because State officials have repeatedly ignored and delayed implementation of a series of broadly-supported plans that would work for all Californians. This predicament has been exacerbated by stakeholders falling short in coalescing a strong-enough coalition to overcome government inaction. The result is a mutually-reinforcing political paralysis and an increasing jeopardy to California's environment and economy.

There would be enough water to go around in most years if the state had sufficient facilities to capture, convey and store a lot more water in wet times than is physically possible today. Hydrographs for the last century show that only about three years out of every 20 are "average" with the balance being either "wet" or "dry." While in many individual years there is not enough water for all needs, when averaged over time there is sufficient supply. Not surprising, the most

Photo: CA Department of Water Resources

conflicts in demands—particularly between fish and farms—occur during times of low rainfall. Further, outflow from the Delta to the ocean varies widely, from as little as 6 million-acre-feet in dry years to 43 million-acre-feet in wet years.

Ironically, more water has been exported out of the Delta during dry times than during wet times, historically because the necessary facilities don't exist to take advantage of an abundance of water in wet periods.

The challenge—and opportunity—is to construct the essential infrastructure to capture significantly more water in wet times, convey it south of the Delta, recharge groundwater basins in the San Joaquin Valley for agriculture, and fill reservoirs in Southern California in order to get through times of low rainfall.

This would leave more water for Northern California and the Delta during dry periods when it is most needed for the fish. This common-sense approach is referred to as "water banking" with a "big gulp-little sip" operating regime—distinctly different from current practices. Of course, there also must be an aggressive commitment to all water-use efficiencies—conservation, recycling, watershed management—linked to new construction projects that is required of all who benefit from such infrastructure. And, there must be investments in a Delta strategic levee system coupled with improved through-Delta conveyance.

A deep and abiding commitment to efficient use of water—which I call a "New Water Ethic"—is the foundation for responsible water resource management and the prerequisite for right-sized facilities. And, then there also must be bold action to build the adequate infrastructure to protect and enhance both the environment and economy in the 21st Century.

The tired debate of "conservation" versus "construction" is a false choice—both are needed.

Thirty years ago, I led the fight against the Peripheral Canal scheme because it would have been an environmental disaster. It was a huge, isolated conveyance facility sized at 21,800 cubic feet per second, capable of taking all the current average fresh water flows in the Sacramento River.

Nobody today in their right mind is proposing an isolated conveyance facility of that enormity. The Governor's current plan for twin tunnels to circumvent the Delta is a combined 9,000 cubic feet per second- just 40 percent of his original proposal in 1980.

Equally important, we launched the successful statewide referendum on the Peripheral Canal because the state would not agree to build the essential storage before operating the "big ditch"—which only deepened suspicions in Northern California about the real motives of the proponents. Without the physical ability to actually capture more water in wet times, an isolated conveyance facility alone is perceived as a "death threat" to the Bay Area environment and economy because of its capacity to divert so much vital fresh water from the Delta ecosystem.

Thirty years ago, I led the fight against the Peripheral Canal. The Governor's current plan for twin tunnels is just 40 percent of his original proposal in 1980.

The essential components and linked actions for responsible water management—conservation and construction, storage and conveyance—have been the foundation of numerous consensus plans: Southern California Water Committee-Committee on Water Policy Consensus Conservation Agreements (1991); CALFED Bay-Delta Plan (1998) and Record of Decision (2000); and Delta Vision Strategic Plan (2008). They also have been embraced in reports from the Delta Stewardship Council, Association of California Water Agencies, and the Governor's State Water Action Plan, thanks to a growing chorus of water leaders and stakeholders statewide calling for a comprehensive fix that includes the key components of an integrated solution. There is no "silver bullet" but there is "silver buckshot" that will get the job done.

Thus, the poetic answer to the question "More or Less?" is "Yes"—and the cosmic reality is that the glass is both half-full and half-empty depending on whether or not we all take responsibility for leadership and action. Now is the time for the State and stakeholders alike to implement a workable solution that will protect our precious environment and support our vital economy to fulfill the promise of the Golden State for all Californians.

Sunne Wright McPeak

A Delta Renewed

Letitia Grenier
Program Director and Senior Scientist
San Francisco Estuary Institute

I hope to see a future Delta with restored areas that support our native fish, birds and other creatures... enhancing the agricultural character, vital water supply function, and unique culture of the Delta.

Born and raised in Southern California and living as an adult in the Bay Area, I am and always have been a user of and dependent on California's water. My personal history has also allowed me to spend countless hours hiking, studying and playing in the state's estuaries, coastal sage scrub, oak woodlands, and beaches. Perhaps this is why I seek a future for our state with healthier natural systems and healthier urban systems – a better quality of life for California's people and wildlife. I see these goals not in conflict with one another, but necessarily intertwined in parallel.

My colleagues at the San Francisco Estuary Institute and I have come up with ideas for large-scale restoration of Delta's natural systems that blend restoration with the working landscape of agriculture, water supply, and urban areas. This approach is based on the collaborative enterprise of science, restoration, and wildlife-friendly agriculture that has been occurring across the Delta for decades, and it adds one new element – a deep understanding of how the historical landscape used to function circa 1800, when it was managed by Native Americans and had not undergone the wholesale transformation we see today.

This understanding of the past nature of the Delta – its vast marshes, complex tidal channels, and extensive and diverse flooding regimes – underlies our approach to thinking about restoration. To gain back some of the desired support for native wild animals and plants that has been lost, we envision re-establishing the landscape-scale physical processes (like beneficial flooding) and habitat configurations that are more like the environment these species evolved in. This landscape can better support native species and it can be more resilient to climate change and other stressors. For example, freshwater tidal marshes are capable of growing peat to rapidly gain elevation, thus keeping up with sea-level rise. Loss of elevation across Delta islands is an ongoing problem that threatens agriculture, water supply, and human settlements with flooding and salt intrusion.

Our approach means taking a large-scale and long-timeframe view of restoration – ecosystems need space and time to function and evolve. It also means using natural processes where appropriate and emulating natural processes with management actions in many cases. A floodplain can be inundated, grow a food web, feed fish and allow them access back into the channels with natural flood flows or through carefully orchestrated water management coordinated among farmers and natural resource managers. The Delta, and Yolo Bypass in particular, offer many hope-inspiring examples of supporting wildlife in an agricultural landscape.

By drawing on our historical understanding of how the Delta functioned, and incorporating new and time-honored approaches to managing water and land to benefit people and wildlife simultaneously, we hope to help design an efficient approach to Delta restoration. When individual projects are designed to add up to a landscape-scale vision, the benefits from each project can add up to more than the sum of the parts – thus maximizing improvements to ecosystem health while making judicious use of the limited land, water, and funding available.

I hope to see a future Delta with restored areas that support our native fish, birds and other creatures and offer people a place to recreate and enjoy nature – all integrated with and enhancing the agricultural character, vital water supply function, and unique culture of the Delta.

Computer generated image of South Delta restoration by San Francisco Estuary Institute

Salmon Migration

Stephanie Taylor

Nothing symbolizes 'made in California' more than the iconic salmon.

Chinook salmon, born in inland watersheds, migrate to the sea to mature and return to where they were born, to reproduce. But their access to natural spawning grounds in the foothills and mountains is prevented by dams on most of California's waterways.

The salmon's fight for survival on the American River exemplifies the consequences of man engineering nature.

While Nimbus and Folsom dams provide hydroelectric power and flood control for us, they block miles and miles of habitat for salmon to lay and fertilize eggs. As populations declined, man turned his attention to rescuing salmon. Completed in 1958, the Nimbus Fish Hatchery creates generations of salmon and steelhead.

Here, migration ends and begins.

When the fall run of Chinook salmon returns, the hatchery erects a barricade across the river. Hundreds of people come to celebrate the heroism of these fish, to marvel at the wonder of nature.

This year, thousands of salmon have fought their way back to fulfill their biological destiny. Against cascading water, salmon enthusiastically leap 20 levels of the fish ladder, in bubbly anticipation of spawning.

These Chinook are native but not wild.

About 6,000 salmon entering the hatchery will produce 4.4 million eggs to be artificially fertilized. Hatchery-bred salmon differ from those born wild, where diversity strengthens genes and the species. While the abundance of salmon is a tribute to the success of hatcheries, some scientists worry about the long-term survival of the species.

Fish ladder below Folsom Dam

Ideally, the salmon are released into the river at the hatchery near Hazel Avenue. In years when the population has crashed, to the edge of extinction, trucking millions of juveniles farther downstream helps ensure their survival. Sometimes they are trucked directly to the San Francisco Bay.

After their release, juveniles swim to the Pacific Ocean, as far as Siberia or Japan, for three to five years before heading home. Shiny silver Chinook, three feet long and 30 pounds, will pass under the Golden Gate, navigate the Delta, swim up the Sacramento River, and into the American. Leaving salt for fresh water, skin darkens, reddens, mottles white with fungus, and fearsome jaws jut.

Downstream from the hatchery, I'm floating in a canoe over newly laid gravel designed by habitat managers to imitate natural sites for salmon to spawn.

Not gentle creatures, they're powerful and swift. They leap from the river and skitter over shallows splashing. Their energy fills the air. They're driven.

Post spawning, they lay gasping and dying. Spent carcasses float and bump gently against the shore, to the sounds of hundreds of crying gulls and vultures waiting to feast.

Mission accomplished.

Salmon, one indicator of the health of our ecology, are what connects the ocean to California's interiors. Vital nutrients pass from sea to soil, from vineyards to forests, across a diverse and challenged system. A vital source of protein, 100 species depend on salmon, from birds to humans to killer whales. And salmon contribute hundreds of millions of dollars to California's economy.

Shiny silver Chinook, three feet long and 30 pounds, will pass under the Golden Gate, navigate the Delta, swim up the Sacramento River, and into the American.

Charles Darwin said that it's not the most intelligent or the strongest species that survives, but the most adaptable.

In our engineered world, salmon are struggling to adapt, and now their very survival is dependent on man.

San Joaquin Valley

Arid

Drought

Oranges

Fallowed land

Environmental justice

Grapes

SWP

Farm profits

California Aqueduct

Almonds

CVP

Water in Food

Water Marketing

San Joaquin Valley

Mysterious Adaptations: Vernal Pools

Stephanie Taylor

What are vernal pools, are they endangered, and why should we care?

Vernal pools are a mysterious phenomena concentrated mostly where the Great Central Valley gently rises to become the foothills of the Sierra Nevada. They're hard to find and they're vanishing. The land looks barren and softly golden as we drive this part of California. But stop, and an undulating countryside transforms into a surreal fantasy-scape.

We most often hear about vernal pools in relation to the "endangered" fairy shrimp. I was always slightly annoyed that such a big deal was made over tiny creatures we can barely see.

In wet seasons, rain collects in shallow depressions of hardpan soil so dense that water can't drain. Every freshwater pool is as unique as a fingerprint, with hundreds of species adapting to precise conditions of its specific environment. An integral element in California's fresh-water system, each vernal habitat is like an island, a microcosm that represents our wider ecosystem of interdependence, including specialists and generalists, each with a talent or trick for survival.

Fleeting vernal habitats are easiest to see in the spring. Flowering plants proliferate in concentric rings, as the water recedes, in an impossible array of color and pattern. When John Muir walked this valley, he said, "Sauntering in any direction my feet would brush about a hundred flowers with every step, as if I were wading in liquid gold." With summer, the bounty adapts through the desiccating heat and in fall, the magic is covered with a sheltering layer of dying matter.

Winter hides what I think are the most fascinating secrets of vernal pools. One chilly and overcast New Year's Day, I spent four hours sitting on the ground by one dry vernal pool, with Dr. Bob Holland, an ecologist who's been passionate about vernal pools for forty years.

On another day, from the top of a gentle hill, hundreds of fuzzy mounds descend in all directions. These are Mima mounds, covered with dead grasses highlighted white in the late afternoon sun. Each casts a shadow over slightly greener, flatter areas that should be filled with water reflecting a huge sky. The pools are dry now, since there's been so little rain.

Vernal habitats are sanctuaries for survival, concealing astounding diversity, innovation and specialist adaptations.

For example, some bees live solitary lives. In spring, each female digs down about 12 inches, and builds natal chambers that radiate like spokes on a wheel. In each chamber, she places one egg on one ball of pollen.

With spring again, tiny, furry adult bees emerge and mate just in time to pollinate their particular choice of flower, dig chambers, make pollen balls, and start the cycle again.

Deep within the muddy pool, the dormant coyote plant sends a snorkel up to the oxygen above, and in spring grows tall, its thistles waiting to do what thistles do in summer. As for fairy shrimp, their ability to lay dormant for as many as one hundred dry years makes them remarkable. All they need is a little rain.

Vernal pools represent a chain of energy that mirrors humans, with decomposers, producers and consumers. They are incubators for an interdependent system of life and food, from expended matter and algae, to coyotes and the majestic Great Egret. One vernal pool can provide a child with a hands-on science lesson that will last a lifetime.

This Valley is one of only five extreme Mediterranean climates in the world, with wet winters and long, dry summers. Out of all the other vernal habitats, California boasts the most spectacularly diverse display, and nobody knows why.

Since John Muir walked this land, 90 percent of the vernal pools have disappeared. What's to be done with the last 10 percent?

Scientists search for rare plants in vernal pool landscape.

Change is Constant

Rita Schmidt Sudman

Without water, California would not be the nation's largest farming state, producing about half of the country's fruits, vegetables and nuts.

Without water, California would not be the nation's largest farming state, producing about half of the country's fruits, vegetables and nuts. Many of these crops are grown in the San Joaquin Valley, the land between Sacramento and the Tehachapi Mountains, where water created abundance but also seeded deep controversies. In the history of the San Joaquin Valley, farmers have both fought change and promoted it.

Epic struggles

After the gold ran out, early settlers turned to farming the rich lands of the Central Valley – from Redding to Bakersfield. There was plenty of flat and fertile land. They "dry land" farmed because there was little developed irrigation. By 1889, California was a great wheat producing state, second only to Minnesota. In fact, the 1901 best-selling novel by Frank Norris, *The Octopus: A Story of California*, was the tale of conflicts between wheat farmers and the railroad barons over land and the price of freight. At that same time, a series of severe droughts hit California causing the collapse of cattle ranching and the failure of wheat crops. When the railroad's new refrigerated cars came on the scene, farmers had incentives to grow and ship fruits and vegetables to eastern markets. Many farmers struggled to bring irrigation to their lands so they could grow these higher priced crops. There were, however, a number of problems related to water rights.

The state had developed a system of dual water rights; rights for land next to a river (riparian rights), and rights for water diverted to lands away from a river (appropriative rights.) Among the earliest lands ranched, with the strongest water rights in the San Joaquin Valley, were the properties of Miller and Lux, two German immigrant butchers from San Francisco. In the 1880s, they bought hundreds of acres that included attached water rights all along the San Joaquin and Kings rivers. They built the Mendota Pool to mingle water and divert it through canals to their lands only. Court battles gave the upper hand to the riparian rights of Miller and Lux.

Valley view from the Tehachapi Mountains

Losing in court was a blow to the appropriators who were trying to create water projects to irrigate Valley lands that were not adjacent to rivers. Worse for appropriators was a 1926 state Supreme Court case ruling that riparians could not be deprived of their right to water, even if they used it wastefully. Public outrage against the riparians led voters to pass a 1928 California constitutional article prohibiting the waste of water and requiring its reasonable use. This "waste and unreasonable use" doctrine remains a powerful tool today because it allows for legal changes based on society's evolving definition of reasonable use. It also removed a big obstacle to federal and state water project planning because projects were based on appropriative rights.

The other problem was climate. Although the San Joaquin Valley contained two-thirds of the Central Valley's irrigable land, it received only a third of the state's precipitation. Wells, powered by the newly invented gasoline and electric pumps, were the obvious answer to the water problem.

Groundwater levels and water quality were different on the two sides of the Valley. The east side of the San Joaquin Valley was the earliest farmed because lands adjacent to Sierra streams contained full groundwater basins. The west side had productive soils, but wells generally had to be drilled deeper, and there were salinity problems because of proximity to the salt and mineral laden Coast Range.

As more land was farmed with groundwater, water levels dropped steadily on both the east and west sides. Meanwhile, there was always the threat of annual flooding.

Mendota Pool, an early diversion point in the San Joaquin Valley

Pine Flat Dam on the Kings River

The great projects define the Valley

The planners and politicians of the time settled on the idea of managing the state's water by building large-scale projects to control flooding and bring water from "areas of excess to areas of need." If a large amount of water was brought into the Valley, the farmers could decrease the use of groundwater that was becoming both unreliable and expensive. However, in the early 1930s, the state did not have the funding to implement a massive plan. Because of the Depression, the federal government was looking for ways to employ workers. So in 1935 the government responded to California's cry for help by providing a New Deal public works project that became the immense Central Valley Project (CVP). All Americans benefit from food produced through that project today.

The epic struggle of Dust Bowl farmers migrating to the San Joaquin Valley to pick the irrigated crops is told in *The Grapes of Wrath* by John Steinbeck. Three decades later, the State Water Project, passed by California voters in 1960, brought additional water to the Valley and opened more land to farming, especially on the west side.

From low-value to high-value

For more than 25 years, I've been traveling up and down the San Joaquin Valley on educational water tours. I've seen firsthand the changes in cropping and irrigation patterns. Twenty five years ago, cotton was king in the Valley and was the top export crop in California.

In 1988, federally price-supported cotton was grown on almost 14 million acres in the San Joaquin Valley. The popular crops now, almonds and pistachios and other nuts, were grown on less than 500,000 acres.

Today very little cotton is grown in California. Other crops that were strong in the 1980s were heavy water using low-value crops, including alfalfa and pasture grasses. The trend to higher value crops began in the early 1990s, when federal water became more expensive due to congressional reforms and environmental regulations. California also suffered periodic droughts, adding to the regulatory limited federal and state water deliveries. Farmers then started using their water for valuable fruits and nuts which increased about 40% between 2000 and 2010. By 2014, these crops made up over 85% of the crop revenue in California.

The trend to higher value crops began in the early 1990s, when federal water became more expensive due to congressional reforms and environmental regulations.

The change to higher-valued crops was long advocated by the environmental community. Starting in the late 1970s, continuing in the 1980s, championed by Marc Reisner, author of *Cadillac Desert*, California farmers were criticized for using enormous amounts of water to grow low-value crops like alfalfa and price-supported cotton. When farmers began to shift to growing higher-valued crops, it was not because of the criticism, but for higher revenues. In an ironic case of "be careful what you wish for," these crops did not free up water for the natural environment as environmentalists had hoped. Instead, the growth of these permanent crops, dependent on water every year, actually "hardened" water needs in the Valley. Farmers could no longer shift crops annually or fallow fields because almonds, pistachios and fruit trees need an annual and consistent water supply.

The genesis of water marketing

Another change came after the 1977 drought with the growth of water markets. In that drought there were no market transfers of water to facilitate moving water around the state from one user to another. When the water marketing concept was developed in the early 1980s by two University of California professors, Dr. Henry Vaux and Dr. Richard Howett, I remember the distrust of the concept from agricultural and urban water users. At that time, agricultural water rates were generally low. Incentives to sell water didn't exist yet. In the San Luis service area of the CVP, agricultural water users in 1990 paid about $16 an acre-foot. Also a popular movie of the day, *Chinatown*, shed a negative light on the idea of selling water rights from one area to another. The movie was based loosely on the Owens Valley history where water rights were purchased surreptitiously by the City of Los Angeles, and the locals lost the opportunity to develop the agricultural economy they desired.

In an ironic case of "be careful what you wish for," the shift to growing higher value crops did not free up water for the environment.

As water prices and incentives grew, the market transfer concept gradually took hold. Today there are sales on the private water market to desperate San Joaquin Valley farmers, who pay up to $2,500 an acre-foot for water that was a couple of hundred dollars only a few years ago. Water was purchased in 2015 from Sacramento Valley rice farmers for about $700 an acre foot and sold within the Westland Water District after delivery for up to $1,300 an acre-foot.

Water marketing is not the entire solution to the San Joaquin Valley's current water woes because all farmers cannot participate. Only those farmers growing annual crops, not almonds and other crops requiring watering each year, can enter into such arrangements. Another challenge to water marketing is the need for more connections between the state, federal and local systems to enable more coordinated management. And then there is the continuing issue of getting water through the California Delta because of the current lack of a Delta conveyance system.

Challenging times

The latest record-breaking drought has challenged the farming culture of the San Joaquin Valley. While some farmers have suffered, others have made record profits on crops like almonds and pistachios. Agriculture in the Valley continues to change. Some Valley agricultural observers predict a 15 -20% reduction in crop acreage over the coming years. They predict some lands will go out of farming due to water and environmental restrictions and the high cost of farming marginal quality lands. Also, efforts at the state and federal levels to undo environmental regulations have been largely unsuccessful.

Today's San Joaquin Valley farmers must innovate, and continue to learn new ways to farm with less water to keep Valley food on our tables.

For those remaining in Valley farming, there is no returning to the days of abundant and inexpensive water. Like earlier generations, today's San Joaquin Valley farmers must learn new ways to farm with less water and to create innovative partnerships between governmental and private interests to keep the San Joaquin Valley's food products on our tables.

Winners or Losers

Tom Birmingham
General Manager,
Westlands Water District

The conversion of millions of acre feet of water, from predominantly agricultural uses to environmental uses, begs the question of what do the people of the state want the state to look like.

The geography and history of California has been shaped by many forces, but perhaps none more than water. From the early days of the Gold Rush, the development of water resources and conflict concerning competing uses of water, have determined how this state would develop. The development of water resources for hydraulic mining began in the early 1850s, but the devastating effects of that use of water on the environment and on agriculture resulted in an early example of the judiciary intervening to protect the environment. In Woodruff v. North Bloomfield Gravel Mining Co. (1884), the court entered an injunction prohibiting the use of hydraulic mining in areas tributary to navigable streams and rivers because of the devastation to the environment and harm to downstream water users caused by the practice.

Contemporary California is also the product of the state's water development. With perhaps the exception of urban areas in the Central Valley, the people and the commerce of almost every major urban area from San Francisco to San Diego depends on the ability to capture water and transport it from distant regions of the state to those urban areas. Indeed, the current residents of California are the beneficiaries of the leadership and foresight of people like Governor George Pardee, William Mulholland, Michael O'Shaughnessy, Governor Edmund G. "Pat" Brown, and A.D. Edmonston. Without construction of the Central Valley Project and the State Water Project, California would not be the most productive agricultural state in the nation.

In 1957, Bulletin No. 3, the California Water Plan, observed "California is presently faced with problems of a highly critical nature – the need for further control, protection, conservation, and distribution of her most vital resource – water." The same is true today. But the problems faced by California today are complicated by a use of water that has developed only over the last 25 years: environmental water use. Not only has the state's population, industry, and agriculture continued to grow over the last several decades, today, more of the state's developed water is managed for environmental uses than either agriculture uses or urban uses.

Absent action to replace water taken from other uses for the environment, the economy of this state will regress.

I do not mean to suggest that the use of water for environmental enhancement is bad; maintaining clean rivers and streams and habitat for native species are important public purposes. But the conversion of millions of acre feet of water from predominantly agricultural uses to environmental uses begs the question of what do the people of the state want the state to look like. Absent action to replace water taken from other uses for the environment, the economy of this state will regress. Although California is experiencing an unprecedented drought, the water supply of large areas of the state, particularly the westside of the San Joaquin Valley, from Tracy to the Tehachapi Mountains, has very little to do with hydrologic conditions. Rather, the water supply of this region is dependent on how much water can be moved from the Sacramento River watershed to the San Joaquin Valley in light of regulations imposed under the Clean Water Act, the Endangered Species Act, and other federal and state laws.

California is a great state because the people of the state have supported water projects that benefit the entire state. Heretofore, no one has suggested that the state can prosper only if someone loses. The real question is whether the leaders of the state have the political courage and will to follow the path of their predecessors.

Thomas W Birmingham

Shown in a drought year, water from the San Luis Reservoir serves Westlands Water District, the largest water district in the U.S.

Water in Food: Your Water Footprint

Rita Schmidt Sudman

Americans consume about 2,000 gallons of water a day, including the water to produce our food, household products and clothes.

Have you ever thought about calculating your water footprint or your use of virtual water? These terms - “water footprint” and “virtual water” - refer to the invisible water embedded in our lives. According to the National Geographic Society, middle-class Americans consume about 2,000 gallons of water a day including the water to produce our food, household products and clothes. Not surprisingly, this amount is much higher than any other country on earth. Studies vary, and numbers can be broadly argued, but the amount of water it takes to provide the American lifestyle is high.

I’ve been interested in this topic for years, even before these terms were invented. More than 20 years ago, the Water Education Foundation and the University of California published a special study called *Water Inputs in California Food Production*. The study examined the amount of water required to produce selected dairy, beef, poultry, grains, soy, fruit and vegetables from the beginning of planting up to the time of cooking. Step-by-step calculations were used to figure the water requirements for each food. There were a number of criteria, including counting all water – whether naturally available or irrigation water. No distinction was made between precipitation, surface water or groundwater.

I thought it was interesting when the figures for water in food were released by the Foundation and the University. Immediately, the figures were controversial. Those who supported increased water development – building traditional canals and dams – said the figures showed more water was needed to produce these foods in California. Critics of water development said the figures showed the large amount of water it took to produce certain foods, in particular beef, proved that people should eat less beef, and generally eat lower on the food chain. The number for beef is high because livestock are fed large quantities of grain and corn over the course of their long lives.

Photo: CA DWR

The Foundation's report said it takes:

- 615 gallons for a 4 ounce hamburger
- 330 gallons to produce 8 ounces of chicken
- 63 gallons to produce one egg
- 8 gallons for one small tomato
- 25 gallons for one ounce of white rice
- 36 gallons for 2 ounces of pasta
- 80 gallons for one ounce of almonds

Studies vary, and numbers can be broadly argued, but the amount of water it takes to provide the American middle-class lifestyle is high.

A powerful industry group disagreed with the study's findings. The California Cattlemen's Association had supported an alternative study citing a much lower figure for the production of beef. That study was based only on water from dams and canals, and included averaging figures with Midwest states where farmers generally did not irrigate. However, gradually the Foundation's water in beef figure became accepted by the agricultural and environmental communities.

More recently, the University of California published another study that analyzed the amount of water used to grow almonds in the state. It reported 1 million gallons of water per acre per year, resulting in the most often cited one-gallon-per-almond figure, even with the increased efficiencies in agricultural irrigation over the last two decades.

Currently, a good source of statistics on our water footprint is the Water Footprint Network, which is the basis of the National Geographic Society's personal water footprint calculator. The Water Footprint Network was based on work in the 1990s by British Professor Tony Allan who was looking at virtual, or embedded, water to understand how arid countries can better feed their people. For his work, Allan won the 2008 Stockholm Water Prize. He based his statistics on U.S. Department of Agriculture's now smaller serving sizes than those cited by the Foundation's earlier water in food study. However, Allan still puts an 8 ounce serving of beef at 1,372 gallons per serving – only about 140 gallons less than the Water Education Foundation's and University's study in the early 1990s.

In a newer study in 2002, Dutch Professor Arjen Hoekstra (working at a United Nations organization, UNESCO,) measured the water footprint along the complete supply chain including measuring the amount of water to produce goods and services. This provided the ability to measure the virtual water of goods that are produced in one location and consumed elsewhere. Hoekstra's estimates account for domestically grown and imported and processed foods. These statistics total all the water used to grow edible crops, to process food and clean up pollution from food production, and also account for the global change in diets over the past 50 years.

As China and India emulate our water intensive diet, there is a greater world-wide demand on water resources.

Americans like to eat and we've been eating more, especially more processed foods, since 1960. We consume twice as much vegetable oil as we did in 1960, much of this due to our increased eating of processed foods. More food means more water used to grow and produce those foods. Our sugar intake, which is of no nutritional value, has risen 14 percent since that time, equivalent to 2,010 gallons of water per year per person.

If your head isn't spinning with numbers yet, consider this.

It is no longer possible to just look at what Americans grow and eat because agriculture is a global business. As countries like China and India emulate our water intensive diet, there is a greater world-wide demand on water resources and more connections between distant countries. Take, for example, the 2013 pork merger between Virginia ham producer Smithfield Foods and Shuanghui, a Chinese company. Critics say the deal allows Smithfield to sell pork to China while Shuanghui effectively gets U.S. water for hog production at the Virginia facility.

The numbers compiled by all these studies during the last 25 years show us that we will all need to consider how much water it takes to fuel our lifestyles.

The Water Education Foundation's "Water Facts" information listing water required for production of food servings and home: www.watereducation.org

The National Geographic Society Water Footprint Calculator: www.environment.nationalgeographic.com/environment/freshwater/change-the-course/water-footprint-calculator

Securing the Human Right to Water

Laurel Firestone
Co-Executive Director and
Co-Founder, Community Water Center

Much of the prejudice and conflict I have seen are because our society remains segregated, and the disparities are so great that we just see each other as 'Other.'

One of my first memories of working on water when I moved to Tulare County after law school was a fistfight breaking out in the middle of a water board meeting. I had often heard the oft-quoted line that "whiskey is for drinking and water is for fighting," but hadn't fully appreciated it until then. Over my last decade of working on community water issues, I've come to appreciate why water is so deeply important to every aspect of people's lives, and also how we can't continue to fight over it if we want anything to change.

From those early days, I also vividly remember being the only woman and the only person under 40 serving on a local water commission. At one meeting, a fellow commissioner publicly announced that I had nothing useful to contribute. We have a long way to go before we see this state's diversity reflected in its water decision-makers. Much of the prejudice and conflict I have seen are because our society remains segregated, and the disparities are so great that we just see each other as "Other."

Californians need to recognize that we are in this together. We can't afford to fight with, pollute, exclude, or steal from each other. Only by embracing California's diversity, ingenuity, and persistence can we come out of this latest drought stronger than we were going in.

Most Californians take for granted that safe water will come out when they turn on their taps at home, work, or school. But that isn't the case for tens of thousands of families, whose water is unsafe to drink or use. This is an entirely preventable and solvable problem. Addressing it will take all of us making some fundamental changes: first and foremost, making it a priority to ensure that all Californians have access to a basic human right – safe, clean, affordable, and accessible water adequate for human consumption, cooking, and sanitation.

One such fundamental change is to protect and restore the primary drinking water sources beneath our feet: local groundwater aquifers. It's hard to care about things you can't see, but remediation and restoration are far more difficult than prevention. Minimizing pesticide and fertilizer application and increasing water use efficiency in agriculture can save money and maintain the long-term viability of both California's agricultural industry and its agricultural communities.

It's hard to care about things you can't see, but remediation and restoration are far more difficult than prevention.

There are many stories of suffering in this latest drought – families borrowing buckets of water from neighbors just to bathe their children and flush their toilets. And many suffer as they did before the drought, enduring unsafe levels of nitrate and arsenic in their tap water. But there are also many stories of people coming together to help their neighbors and find solutions. It is these efforts that will shape our path out of this crisis.

Drought has horrible and acute impacts on the lives of many of our most vulnerable families. But drought also has a silver lining. We can use this crisis to change the way we protect and provide drinking water, thereby creating more resilient communities and ensuring the human right to water for all Californians. And we will only get there if we resist the urge to fight each other and instead come together around this crisis.

Laurel Firestone

This private well is more susceptible to groundwater contamination than public wells that are tested regularly.

A local resident fills buckets of donated water because the well in herCentral Valley home has no water.

San Joaquin Valley

River Restoration: Native plants

Stephanie Taylor

The results demonstrate that native plants can dominate non-native plants that squander water, and that rivers can flow naturally, given space, within the confines of engineering.

Riding high in the cab of a mowing machine, the farmer drives slowly. Baby quail run for their lives. His aim is to harvest native plants for seed, he tells me as he gives creatures time to flee from the spinning blades.

The farmer is a pioneer in native plant restoration with years of methodical cultivation, matching seed from specific eco-regions to compatible sites in California. His land is a living laboratory for what he preaches, from hedgerows of rushes and to dune sedges to saltgrass, lupine and clover. He develops seed by selecting the best attributes of each species, rather than modifying them genetically, to return to fields and watersheds for weed and erosion control, water quality and habitat restoration.

What's so important about native plants when almost all of the plants in California are non-native, all introduced since Europeans arrived? One answer: it's hot and it's getting hotter. Drought tolerant native plants have thousands of years of adaption to the extreme conditions that exist here. Second answer: riparian plant species help Central Valley rivers flow naturally, critical to flood control and habitat preservation within our over-engineered landscapes.

The farmer walks into a field and wanders between rows of fuzzy milkweed and grasses. He picks a sample of grasses and rolls the seed in the palm of his weathered hand. He inspects plants for destructive invaders. A monarch lands on a milkweed, ignoring iridescent green beetles chewing. Sixty native species in rows of grasses and plants wait for harvest, seed separation, propagation and delivery. I ask him where the seeds and young plants go next. Restoration projects from the northern reaches of the Central Valley and farther south benefit from decades of experimentation.

Restoration site on the San Joaquin River

A meandering river is a healthy river and one that nurtures a diverse ecosystem along and within its banks.

At the confluence of the San Joaquin and Tuolumne rivers, from a bluff overlooking other restoration sites, sandbar willow, black willow, creeping rye, evening primrose, cottonwood, box elder and valley oak flourish. Here, landowners are encouraged to remove old levees, to let the river meander. This is where the farmer's seeds and seedlings have come to grow. In a field amended with byproducts from nearby canneries, they prosper, selected to adapt to the area's frequent flooding.

In partnership with nature, purposeful restoration can mitigate harm we've done on the human side.

Rivers and wetlands in the Central Valley are all about flood plain physics -- velocity, sediment and diversity of wildlife habitat. State, federal and private entities collaborate all over California. The results of this science demonstrates that native plants can dominate non-native plants that squander water, and that rivers can flow naturally, given space, within the confines of engineering. In partnership with nature, purposeful restoration can mitigate harm we've done on the human side.

Groundwater

Wells

Drip Irrigation

Canals

Subsidence

Almonds

Recharge

Management

Overdraft

Drought

Groundwater

San Joaquin: Valley Views

Stephanie Taylor

At what was Tulare Lake, I try to imagine the thousands of white pelicans that used to nest here, stretching their 10-foot wing-spans in flight on the Pacific Flyway.

From the east side of the San Joaquin Valley, the Sierra foothills offer a dramatic view of changing terrain – from lush to parched. From citrus groves and almond orchards, the landscape shifts from a verdant Tuscany landscape to golden-brown along the Coast Range.

In Lemon Cove and Orange Cove, orchards of lemons and oranges and avocado creep into every fertile niche. Over gently rolling hillsides, patterns of mature orchards merge with new ones. Remnants of winter-harvested navel oranges lie abandoned in one orchard. An imaginary scent of citrus lingers in the air deep within the rows. Across a canal, another promises sweet summer Valencia oranges.

Farmers settled the east side of the San Joaquin Valley first, lured by abundant water from five major rivers and nutrient rich soils. Reservoirs and dams, built to control and capture torrential Sierra Nevada watershed rains, protect against devastating and periodic flooding, and also store water for summer use.

In contrast, on the west side of the Valley, some land is fallowed, some saved for only the most valuable crops like grapes, almonds and other nuts. Row after row of vines stretch endlessly in late afternoon sun, waiting for tiny buds. Fields and orchards near Tupman, for example, are barren, withered. Other towns look dusty, forlorn, graffitied walls crumbling, leaning.

Land was always cheaper on the west side. Until federal and state aqueducts brought water, farmers and communities depended on groundwater. With surface water deliveries reduced, wells had to penetrate deeper and deeper. As more and more water was pumped from the ground, the soil collapsed, subsided.

Subsidence in the Valley can be measured in feet- many feet, with visible damage to aqueducts and bridges. Near the San Joaquin River, a closer look under a bridge displays

On the west side of the Valley, pilings have pulled away from this bridge due to land subsidence caused by severe groundwater pumping

evidence of the impact of subsidence on infrastructure. Pilings no longer support the bridge. New pilings have been constructed to make the structure safe.

At one site near Mendota, land subsided about 30 feet between 1925 and 1977.

I look down at dry weeds and ground under my feet, and try to imagine what another 30 feet down might look like.

This flat land is graced with tumbleweeds, and I've often wondered where they come from. Turns out, they're an invasive species of thistle from Russia.

The Tulare Basin is the states most heavily subsided area. It was named for the famous Valley tules, as in Tule Fog. I see no tules now as the land was drained for farming long ago.

Areas of ground are white with evaporated salt. You can see a salt bathtub ring on Google Earth, from what used to be the shores of a 14,000 acre inland freshwater lake, drained, empty and dry.

East side soil, granitic from the Sierra: west side soil, salty from the sea.

I try to imagine the thousands of white pelicans that used to nest here, stretching their 10-foot wing-spans in flight on the Pacific Flyway. In wet years, Chinook salmon arrived via the San Joaquin River.

This used to be the home of around 70,000 Yokut people. Until Europeans arrived.

A wetland habitat has been created to coax birds from building nests in more saline toxic areas. Mile-long ponds seem to disappear into the horizon. Reflected in shallow water, an anxious avocet dances, darts, hoping to lure humans away from two spotted eggs camouflaged in dirt. They're his legacy, and he's concerned.

The avocet fears humans, but in fact, he has more to fear from natural toxins found in the nearby Coast Range.

Crisscrossing the Valley, from floor to foothills, from salty flats to rich soils, reveals a portrait of how water has shaped contemporary California.

Below: An Avocet in the Tulare Basin   Opposite: Oranges on the east side of the San Joaquin Valley

Groundwater - One Resource

Rita Schmidt Sudman

Like a bank account, we must manage this valuable resource wisely so the investment grows and does not become worthless.

Groundwater is by far California's largest reservoir, holding an estimated 20 times the amount of water stored behind dams. The more than 16 million acre-feet of groundwater that is pumped annually in California is more than the yield of the state and federal water projects combined. Farmers rely on this resource especially in times of drought. Most Central Valley cities and half of all Southern California cities also use groundwater for drinking water. In a normal year, groundwater supplies about 35% of California's total water needs. In the 2015 drought year, best estimates are that groundwater currently supplied 75%, an amount not reached since the great drought of 1977.

Groundwater levels in California are declining and this decline will limit its future use. Also, as extensive groundwater pumping continues, the land is sinking rapidly in areas of the San Joaquin Valley and this puts infrastructure at great risk of costly damage.

Legal confusion

In California, two legal entities govern groundwater use. Since 1914, based on a California Supreme Court decision in 1899, groundwater has been described as being of two types – that which flows in subterranean streams in "known and definite channels" and water classified as "percolating groundwater." I've heard legal scholars argue that these definitions are based on knowledge of geology as it existed 100 years ago. While much of the state's surface water is managed through a system of permits, landowners have been able to pump what they want while using it for a "beneficial use."

The definition of beneficial use can be interpreted by courts to reflect societal changes. By the 1940s, as groundwater depletion grew in rapidly growing Southern California, legal fights led to court-ordered adjudications. When a basin is adjudicated, the courts require cooperation of all pumpers in the basin and appoint a "watermaster" to ensure pumping limitations are complied with.

I've heard legal scholars argue that groundwater definitions are based on knowledge of geology as it existed 100 years ago.

No basins are adjudicated in the Central Valley. Overdraft is a serious problem in the San Joaquin Valley, especially the southern part. Overdraft is simply taking more water out of a basin than the amount of water that is naturally or artificially replaced. It's just like a bank account. You can't take out more than you put in or you will be overdrafted. Like a bank account, resource managers say we must manage this valuable resource wisely so the investment grows and does not dwindle and become worthless.

Overdraft is destructive

With overdraft comes subsidence. Subsidence is the sinking of the land surface due to overdraft. The results of subsidence can be expensive and dangerous, including cracks in irrigation canals and ditches, uneven roads, unsafe bridges, sinking house foundations and collapsing wells. In the San Joaquin Valley, I've seen evidence of subsidence - the gas station with the cracked foundation, the bridge that had to have new pillars to make it structurally sound, and the cracked canals. I've been to the Valley site of the famous 1977 picture of the United States Geological Survey's Joe Poland standing by the telephone pole, marked with the elevations where the land surface used to be in 1925, 1955 and finally 1977. In 2015, the land at this same location was much lower.

In 1979, the California Department of Water Resources (DWR) estimated there were 1.5 million acre-feet of groundwater overdrafted annually. In 1995, DWR estimated that same amount was overdrafted. And recently DWR estimated the same amount again.

How can the figure be similar through the years when surface elevations are lowering and farmers are drilling more and deeper wells?

The answer is that, until a law was passed in 2009, there were no recorded measurements of groundwater elevations and depletions. We don't really know how much groundwater is used because, until 2014, California had no mandatory groundwater management law.

The Central Valley Project (CVP) was built starting in the 1940s, in part because farmers in the San Joaquin Valley were relying on overdrafted groundwater basins that produced increasingly less and poorer quality water. The CVP and the later State Water Project (SWP) water deliveries did relieve the reliance on groundwater for a number of years until drought years, followed by environmental regulations, restricted the delivery of project water, especially to the San Joaquin Valley.

Adding more water back into the groundwater basin is part of the solution. Called recharge, nature does it in generous water years and water managers do it by creating water banks above the basins to recharge them. The Kern Water Bank, southwest of Bakersfield, covers 32 square miles and can recharge 550,000 acre-feet (the size of Lake Isabella above Bakersfield) a year. Half that water can be pulled out each year - if managers have the water available to put in the bank.

Management is necessary

There have been efforts for decades to pass groundwater management laws in California. Notably in 1978, Governor Jerry Brown's Commission to Review California Water Rights Law recommended comprehensive legal changes to managing the state's groundwater. Agricultural interests were strongly opposed and the proposed changes were shelved. California remained one of the few Western states, and finally the only one, allowing landowners to pump without restrictions. The recent severe drought changed the equation. As the groundwater levels sank, some farmers began to believe it was now in their best interest to define groundwater rights, and be sure that the resource was used in a sustainable fashion.

California's first mandatory groundwater management law, the Sustainable Groundwater Management Act, was passed and signed by Governor Jerry Brown in 2014, and became effective in 2015.

Do we have enough time to save this resource for future generations?

Eventually the law will regulate groundwater extraction to ensure a sustainable resource over the long-term, require measurement of pumping, and division of rights to users. Groundwater basins must be in balance by 2040. Farmers are to create working groups and begin making rules leading to a basin in which the groundwater resource is sustainable.

The main idea is to figure out how much groundwater can be withdrawn annually without causing an undesirable result. In 2015, the heavy groundwater pumping continued. In fact, that year the state Department of Water Resources released a startling NASA report saying land in the San Joaquin Valley was sinking nearly two inches a month, and as much as 13 inches in the last eight months of 2014, in the southern part of the Valley around the Tulare Lake Basin. These are new record lows - up to 100 feet lower than previous records.

With the passage of the groundwater law, there is concern about a new "race to the pumphouse" as some farmers may pump more to establish a benchmark amount of water use. Altogether, the plan to create sustainable groundwater basins will take almost 40 years.

Do we have that much time to save this resource for future generations?

Surface and groundwater are one resource and must be managed together.

Around 2000, there was debate whether the word describing this resource should remain written as two words (ground water) or become one word (groundwater). Lots of us working in water at the time weighed in during the lighthearted argument that ensued and the one word usage prevailed. That was a wise decision. Surface and groundwater are one resource and must be managed together.

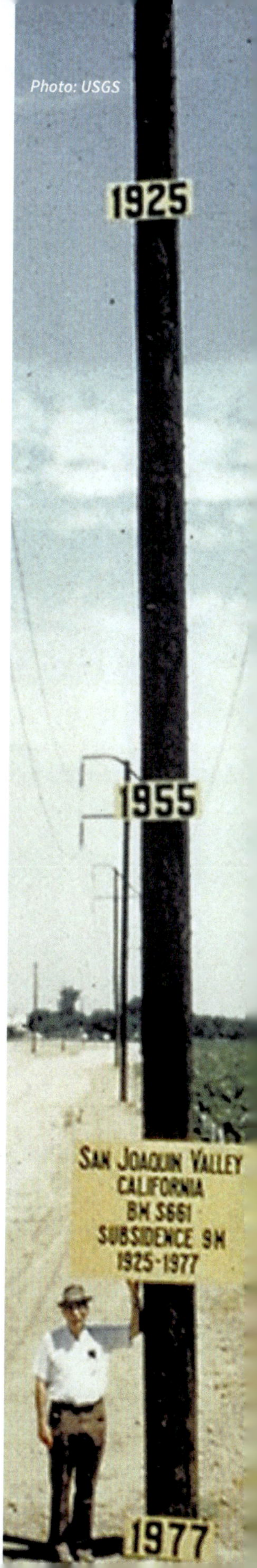

Photo: USGS

The famous 1977 picture of the USGS's Joe Poland, standing by a power pole in the Mendota area of the San Joaquin Valley, marked with the elevations where the land surface actually was in 1925, 1955 and finally 1977. By 2015, the land at this same location was sinking as much as a foot a year.

After More than 100 Years: Plans to Manage Groundwater

Carl Hauge
Chief Hydrogeologist, California Department of Water Resources (retired)

The 1913 legislation that set permits for surface water, but not groundwater, led to a 'wild west' approach to groundwater management.

When the California Water Commission submitted the Water Commission Act to the California State Legislature in 1913, the Act required state permitting jurisdiction over diversions of surface water and groundwater. Thus, it appears that the Commission, appointed by the Legislature in 1911, understood that groundwater and surface water are one and the same resource.

However, a lobbyist from the San Joaquin Valley argued before the Commission that the courts had already established judicial control of groundwater in the 1800s and early 1900s, and therefore the Legislature had no authority to establish laws dealing with groundwater. The lobbyist apparently convinced the legislature because when the Water Commission Act was introduced in the Legislature, it required permitting by the state for surface water diversions only. Groundwater was not included. While this apparently politically motivated and short-sighted legislation was not the cause of California's "wild west" approach to use of groundwater, it did nothing to help develop sustainable management of water resources.

Whether groundwater or surface water is used, the other is always affected. "Separation" into groundwater and surface water is an artificial distinction that has perpetuated the incorrect belief that these are two different sources of water. Both are derived from the same source—surface water flowing into and through the basin. Until this relationship is understood and incorporated into water management plans there is little hope for development of a water budget that will lead to sustainable management of water resources.

If groundwater had been included, the legislation would have applied only to diversions of groundwater off a landowner's property. The California Supreme Court had already declared that a landowner had a right to pump as much groundwater as he could put to beneficial use on his property, but that right was correlative with the rights of all other landowners within the basin. A correlative right is quantified only after the pumpers in a basin have initiated a court action that determines the amount of water that each landowner is entitled to each year, depending on the amount of water in the aquifer. But there is no record of a local agency adopting correlative rights pumping limitations without court action. This is because the directors of many local water agencies are local groundwater pumping landowners so they have little interest in limiting their pumping.

Groundwater has always been used as a "back up" water supply when surface water is not available. Too often, because of the lack of surface water to irrigate and to recharge the aquifer, this has resulted in rapid drawdown of groundwater levels within the aquifers. When groundwater is pumped out of an aquifer, empty space is left in the aquifer. If surface water is available, it flows into the empty space in the aquifer. But if no surface water is available for recharge and pumping from the aquifer continues unabated, groundwater levels will decline resulting in water quality degradation, land surface subsidence, stream flow depletion, and shallow stranded wells.

Directors of many local water agencies are also groundwater pumping landowners, so they have little interest in limiting their pumping.

Although the Legislature had several opportunities to do so, it never amended the Water Code to authorize state agencies to get involved directly in groundwater management. As a result, state agencies claimed that management of groundwater was a local responsibility. Some local agencies have met that responsibility, but too many have not. Dwindling groundwater supplies resulting from a prolonged drought finally spurred the Legislature to enact the Sustainable Groundwater Management Act, signed into law by Governor Brown in 2014. The

Buckling of the Delta-Mendota Canal due to land subsidence

Photo: USGS

Act requires that water resources within each groundwater basin and subbasin be managed so that the resource is sustainable by the long-term.

Local water agencies within each basin or subbasin must form Groundwater Sustainability Agencies (GSA) that must develop a sustainability plan. The plans for certain prioritized basins that are critically overdrafted must be submitted in 2020. Plans for prioritized basins that are not critically overdrafted are required by 2022. The Act then stipulates that GSAs must submit a report every five years that identifies the actions taken by the agency that will lead to sustainable management within 20 years after implementation of the plan.

Groundwater pumping

Photo: CA Department of Water Resources.

The Sustainable Groundwater Management Act requires that water resources be managed sustainably 20 years after the plan is adopted, but provides for certain time extensions. The rapidity with which GSAs are formed and basin plans are adopted will determine which basins meet the Act's deadlines. Early indications are that local water agencies in some basins are moving toward a solution, but others are exhibiting resistance.

Groundwater sustainability depends on successful implementation of the 2014 Act.

The political fiefdoms that have been established and guarded protectively by the landowner-directors of more than a thousand local agencies, and the tendency to treat groundwater as their own separate water source with no regard for the correlative rights of other pumpers in the basin, are both significant obstacles that must be overcome.

After more than 100 years of treating groundwater as an unmanaged commons, the development of groundwater budgets, and a sustainable supply of surface water and groundwater depends on successful implementation of the Sustainable Groundwater Management Act. Success will be achieved only when the water supply is managed to eliminate overdraft and land surface subsidence, and to regulate water quality degradation, and stream flow depletion.

All of these factors must be dealt with if the goals of the Act are to be attained by 2040. The countdown has begun.

Carl Hauge

Water Solutions Require Courage and Collaboration

Dave Orth
Member, California Water Commission,
Former General Manager, Kings River
Conservation District

California often is locked in battles between 'haves' and 'have nots.' We must change that dynamic.

I entered California's water world in 1986 as Westlands Water District's finance director, and began a remarkable journey through myriad difficulties and opportunities. When the Westlands Board invited me to serve as general manager in 1995, I hoped my efforts would contribute to change in how water supplies are managed. In my position at Kings River Conservation District, that effort continued. Along the way, I learned that leadership and collaboration are keys to successful outcomes.

California often is locked in battles between "haves" and "have nots." Progress is defined by taking or reducing water managed by others. We must change that dynamic. Significant advances are made when solutions provide opportunities for all interests to improve.

How do we shift from refighting the same battles to pursuing a well-devised plan of solutions? My experience suggests three things: a desire to build relationships; a commitment to collaboration; and a determination to find common ground.

The Kings River Fishery Program is a classic example of all three. Through collaboration, fly fishermen, river enthusiasts and farmers came together to develop a unique program that creates a sustainable fishery, while providing recreational benefits and irrigation water supplies. Other efforts I have participated in include the Kings River Integrated Regional Water Management Plan, which is readily used as a statewide model for regional planning, and the Governor's Drinking Water Stakeholders Group, which provided practical, implementable recommendations to the governor addressing disadvantaged community water supply challenges.

Creation of the 2014 Sustainable Groundwater Management Act also relied on these three components. I invested myself personally and professionally into developing the Act and I believe it will prove to be a key factor in California's long-term water success.

Drought is a California reality, followed by floods, which will be followed by more droughts.

Now, as a member of the California Water Commission, there is an opportunity to invest in water storage, critical to California's future. As the Commission seeks out beneficial projects for funding, success will be found when all participants embrace a vision of a stronger California and work collaboratively to improve all interests. The reality is that any large storage component still is decades from completion. Storage is needed, but storage is not the "silver bullet."

California's current historic drought has caught our attention. Yet, we all know that drought is a California reality, and this drought will be followed by floods, which will be followed by more droughts. Instead of zeroing in on drought, we should recognize it as a catalyst to change the way we think and the way we use water in recognition of this pattern.

Solutions are developing, and none are easy. Problems aren't solved by restricting land use or telling farmers what to grow. We will reach solutions by collaborating and cooperating to manage demand through carefully managing land use, improving surface water management, and by meeting groundwater sustainability objectives. California's diverse water users can make good decisions if we commit to embrace these actions.

If we can't, we will keep racing to the bottom of our available water supply and the face of California will change.

Almonds: Show Me the Money

Rita Schmidt Sudman

The little nut is not the problem. It's the increased groundwater it takes to produce the nut.

Almond orchards have been called "ground zero" in the latest California water wars because of their water usage. We've all heard the one gallon per almond figure. Despite the current drought and water uncertainty, almonds are so extremely profitable- at up to $6,000 per harvested acre—that 77% of the state's farmers say they will plant them, according to a U. S. Department of Agriculture 2014 poll.

In 2016, selling almonds was a $5.3 billion-a-year business. Almonds now account for about 14% of California's irrigated farmland – about 1 million acres – using about 10% of the state agricultural water. It is the state's biggest export crop, with farmers sending about two-thirds of the nation's almonds to the world market. California's soil and climate make it one of the few places in the world where almonds grow abundantly. Spain is the next highest almond-producing country, but is far below California in exports.

Many world cultures, especially the Chinese, love the high protein nut and are eating 1,000% more California almonds than they ate a decade ago. Almonds are not a junk food, have a long shelf life of eight months, and fit our popular American culture as a gluten-free health snack. The California crop has tripled in production in the last 15 years and in 2016 made up 81% of the state's global supply. By 2013 almonds had become more valuable than the state's wine grapes! In 2014, almond farmers got over $4 a pound, twice the 2011 price. The 2016 prices were slightly lower at about $3 a pound. Nevertheless, farmers say because of these generally increased profits their employees have benefited by much higher wages. The California Almond Board says planting more almonds for profit is an example of the American economic system at its best.

The main problem with almonds is not the little nut. It's the increased groundwater it takes to produce the nut.

Harvested almonds in the Central Valley

Most of the planting in recent years has occurred on the west side of the San Joaquin Valley – the area now with the least water from rivers and dams. Most of the new almond crops depend on groundwater, a finite resource being depleted in many parts of the Valley. About 48,000 new acres of almonds were planted in the Valley in 2014 – 270,000 acres of almonds over the last decade. Other high value permanent crops are being grown, also with groundwater. Pistachio production has doubled. Mandarin oranges have more than tripled. Walnuts are another crop on the rise.

California's entire almond crop takes 1 million gallons of water per acre per year, about 4 acre feet a year per tree according to University of California researchers. The researchers say that almonds use one-and-a-half times as much water as strawberries or tomatoes. However, that's less water than alfalfa, which uses more water in total than any other crop including almonds.

Some industries have used the almond water statistics to show how little their industry's water use is compared to growing almonds. The entire craft beer industry says they use the same amount of water to produce all their beer as farmers use irrigating just 680 acres of almond trees.

If we think about eating lower on the food chain, all nuts, fruits and vegetables take a lot less water to produce than beef and chicken. Various studies, including one sponsored by the University of California and the Water Education Foundation, found that a 4 ounce hamburger took 615 gallons to produce and it took three times as much water, 1,231 gallons, to produce an 8 ounce steak.

Drip irrigation improves almond crop yields but does not recharge groundwater like the old inefficient flood irrigation.

If the available surface water and groundwater is not enough for their high value nut crops, many farmers are willing to buy expensive water for their trees that are permanent crops that require year-round irrigation. For example, by the time water is delivered, some almond farmers around Fresno have paid $2,500 an acre-foot.

Increased water prices in the last two decades have driven farmers to convert to efficient drip irrigation. All the orchards I saw in 2015 in the San Joaquin Valley were on micro drip irrigation. The systems are expensive but these orchards are a 25-year investment. In Westlands Water District, the biggest water district in the country, 93% of the farmers use drip irrigation producing an 85% efficiency rate. Farmers say drip improves their crop yields, enhancing their bottom line. The downside is that drip does not recharge groundwater basins like the old inefficient flood irrigation systems.

In the Valley in 2015, I also saw farmers ripping out melons to plant almonds. Observers say those melons, along with the broccoli and carrots that are not being planted, will make those vegetables scarcer and more expensive. If irrigated acreage in California declines up to 25% as some experts predict, only the most valuable crops such as nuts and grapes will be farmed.

If irrigated acreage in California declines up to 20% as some predict, only the most valuable crops such as nuts and grapes will be farmed.

Critics of the vast almond production and its water use accuse the almond farmers of taking more groundwater than can be replenished – actually mining deep groundwater that took eons of time to accumulate. They say that by exporting all those almonds, we are indirectly shipping our precious water overseas. They call for a state ban on planting new permanent crops – a ban primarily aimed at almonds.

Almond supporters say the idea of dictating to farmers what to plant is against our free market system. They note regulating crops never worked when it was tried by centralized nations, such as communist countries. Farmers also say they have responded to earlier criticism of growing low-value crops, primarily alfalfa and grains, and are now using the water for a higher value crop with healthier benefits to humans.

A few years ago, an almond industry campaign asked consumers to eat more almonds, using the slogan, "A can a week, that's all we ask." It looks like the world is now doing more than its part.

Almonds on efficient drip irrigation

Water and Oil

Stephanie Taylor

Someday, oil production might become a necessary side effect of producing water.

Bakersfield. Two pumpjacks work endlessly on the banks of the Kern River, near a bridge that spans no water. I hear their forlorn cry in the dry October landscape. The larger of the the two emits a soft whistle, coming and going on the wind. I listen to its rhythm, purse my lips together to mimic the exact note on the scale. It lasts about 4 seconds, fading in then out.

Around and around, two arms drive a monstrous head of steel, dropping then lifting a 2" diameter pipe. It sucks oil, sand, and water from a casing drilled far into the earth, sometimes as deep as 20,000 feet.

The second pump emits a low squeak, something unlubricated, exhausted. It's the older of the two, paint faded, rusting steel exposed. An ancient belt-driven generator sounding like an old refrigerator, its pipe sucking and dripping oil, shiny and black. These almost creature-like machines dot the landscape surrounding Bakersfield, past Oildale to the east, and to the Petroleum Highway (33), to the west. Nothing of much value except oil. And water.

About 210,000 wells have been drilled in California, and around 100,000 are active, operated by about 570 companies, large and small. Kern County, the third largest oil county in the country, produces 80 % of all California oil.

This is a story about oil production and water in the southern San Joaquin Valley, and specifically in Kern County where oil and water is defined by millions of years of geology unique to this area. The creation of mountains, sediment, rock and soil, and the migration of fluids under pressure. Diatomite formations: rock formed from the skeletal remains of microscopic diatoms. Sandstone formations: up to 1 million times more permeable than shale. Shale formations: densely packed layers. It's the diatomite that makes Kern County uniquely productive.

By necessity, this is a simplified story about stupefying complexities.

> *There is almost no oil without water. Organic material, buried deep and heated by the earth breaks down to form oil and then natural gas. Permeability and porosity of each unique formation determines how easily fluids move through rock and up to the surface.*

Before Interstate 5 was finished in 1979, driving to LA on 99 was California's blast-from-the-past. I never got off the freeway. Miles of cotton. Tumbleweeds rolled across the highway, spinning with the wind. They foreshadowed what this region used to be, and what it was to become. And the in-between years? The State Water Project brought water to the San Joaquin Valley.

> *Yokuts Indians were the first to use natural seeps of tar near the Kern River in Bakersfield. Commercially viable petroleum deposits were discovered in 1899. By 1903, California was the top oil-producing state in the nation, with most of it coming from deposits in the San Joaquin.*

18 years in LA, I had little consciousness of oil, not even while driving La Cienega over hills graced with nothing but pumpjacks. Saber-toothed tigers at bubbling La Brea tarpits, pumps camouflaged in lush Beverly Hills gardens, I lived at the beach just south of a huge Shell refinery. Never gave oil a thought. All this oil. How was it created, how is it extracted, and how does it affect water?

> *California oil is thick, thicker than most places in the world. Almost a solid, it's like molasses in a freezer – until heated with water injected as steam.*

Bridge over Kern River in Bakerfield after a dry summer.

To the west, the rain shadow side of the Coast Range, rainless hills are so barren that nothing appears to grow, scratched clean. Soil is dry, salty, with geologic names like Kreyenhagen Formation, Monterey Shale, Corcoran Clay. Little is left for the wind to ruffle. Villages almost too non-descript to name, are named – “Dustin Acres” and “Valley Acres.” Roadside signs praise the Lord and various churches. A “Nothing but Jesus” banner graces the side of a metal storage building.

The Coast Range on the west side of the Great Central Valley, used to lie deep within the ocean, uplifted, bringing diatomite rock, salt and minerals, including selenium.

This is a living desert of mechanical creatures, poking the earth, conveying fluids. These are some of the top oil producing fields in the United States. Fields called Buena Vista, Midway Sunset, Elk Hills, Belgian Anticline, San Emidio Nose. Vertical and horizontal structures, pumpjacks and pipes shiny in the sun, transport oil, gas and water 24/7. It appears as a landscape devoid of people, there are so few about.

Deep within rock, oil moves through remnants of tiny creatures, cracks and pores, winding, meandering. Well casing serves as a straw, releasing pressure held within earth for millions of years. In 1910, at 2,440 feet, the famous Lakeview Gusher, on the Petroleum Highway, erupted uncontrollably for 18 months straight, “a river of crude…the greatest oil well the country has ever known.” It spewed about 9 million barrels, about twice as much as the BP Deepwater Horizon spill in 2010. Even today, engineering predictions are “best guesses.”

These are mature fields. Portions have been depleted. Wells remain productive, but not without the stimulation of impermeable rock with enhanced recovery methods. Fracking.

Fracking.

A fractured subject, emotional and contentious. Opinions versus facts, pros and cons. Where does the water go and does it contaminate potable water above and below the surface?

Fracking is "an optimization problem as opposed to an exploration and discovery problem." It's been used on the west side for about 50 years. Usually a one-time preparation event, often lasting less than an hour out of an entire life-span of a well. Technologies developed in the last ten years have increased productivity.

Frac fluids are injected under pressure down wells until nearby rock cracks. Many "frac jobs" on the westside of the valley are "shallow play" vertical wells, around 1,500 feet deep. A few curve and "run" horizontally for one to three miles, "chasing the formation" into oil-rich rock. Holes are blasted through steel casing. A jelly-like substance – around 70% water, 25% sand, and 5% common food additives such as guar gum, and anti-bacterial household chemicals – is injected under pressure, making and enlarging cracks.

Fracking the typical California well uses about 120,000 gallons of water to ready a single well for production. That's about 1/6th the volume of an Olympic size pool, and far less than the average golf course. It's much less water intensive than fracking in other states because California frac wells are typically shorter: 1,500 linear feet compared to 4 miles. Texas wells, for example, can use between two and ten million gallons of water.

Fracking makes cracks about 1 inch by 200 feet wide and long. The cracks provide artificially enhanced permeability to allow oil and gas to move easily, out of rock and into wells. Wells lift fluids to the surface.

At the well head, oil is separated from sand and water. At this point, extracted water might be clean enough to be treated and recycled – either returned to the production process, returned deep into the earth, or sold to farmers to grow crops.

Wastewater is hazardous if not managed properly. It's the duty of California's State Water Resources Control Board and the Division of Oil, Gas and Geothermal Resources to prevent disposal of any contaminated oilfield wastewater near underground drinking supplies.

Or – do contaminates like natural salts on the westside, render the wastewater too toxic to recycle? Under strict regulation and conscientious monitoring, it may be reinjected into deep formations of already salty and unpotable groundwater, as determined by the Safe Drinking Water Act.

> *Innovation is the future. In one year, a company sold around 8 billion gallons of treated wastewater to almond and pistachio growers. Companies plan on increasing transfers to growers, using the latest cleaning technologies, like electric pulsing, to purify wastewater.*
>
> *Could it be that someday priorities will switch, as one expert speculated: that the technological production of oil might become "a necessary side effect of producing water?"*

When Interstate 5 first opened, straddling the foothills of the Coast Range, the view at night revealed little but darkness between small towns in the San Joaquin. Over time, lights filled the empty spaces between towns until it seemed to be one long town from Bakersfield to Sacramento.

> *Dr. Mark Zoback, Professor of Geophysics at Stanford University predicts that, "By 2040, the world is going to need about 50% more energy than is consumed today."*

I sit in my car next to the oil fields thinking about all the petroleum products we've come to depend upon. I remember Dustin Hoffman in his 1967 iconic role in "The Graduate." His dad's friend drapes an arm across his shoulder, and pontificates about the future. "I just what to say one word to you. Just one word. Are you listening?" he asks, "Plastics. There's a great future in plastics."

Protecting Groundwater: Quality and Quantity

Thomas Harter
Professor and Cooperative Extension Specialist, Groundwater Hydrology
University of California, Davis

A Poem

A dream job.
That's what keeps me going,
Twenty years into it.
One leg
In advising and technical consulting
(In my case they call it
UC Cooperative Extension),
Informing people and shaping good decisions,
The other leg in academic applied research.
And no bosses to tell me
Which project to take,
Whom to advise,
Or how the advising and
Outreach must look like.
Instead, colleagues and collaborators inside
And outside the university
That support my work or
Whose work I support.
A lively and diverse and colorful network
For which I care and by which
I feel cared for, even when we do not agree.
The opportunity to think
Outside the box,
To tackle water problems
That matter to many people
Now and in the future.
A university that has held me
Accountable in ways few universities do
(Through their merit system)
Without being prescriptive.
And a public that values the work
We do, I do, at the university.

Over those 20 years
California has begun
A groundwater revolution.
The roots of which lie in the
Invention of the
Turbine pump and of the
Haber-Bosch and other

Chemical manufacturing processes

A century ago.

The seeds of which lie in the

Droughts and

Polluted watershed stories of the

20th century.

The sprouts of which were the

1914 founding of the water commission

(A revolution at the time), the

Porter-Cologne clean water act of California

(Another revolution)

And the Superfund site cleanups following

A Civil Action and Erin Brockovich.

The revolution that I have witnessed,

With my work focused on

The groundwater-agriculture interface of California,

Evolved from the 2002 sunset of

A thirty-year old

Permit exemption for

Agriculture in the Porter-Cologne

Water Quality Control Act,

And the 2014 Sustainable Groundwater Management Act.

Almost overnight, it seems

(As it always does with revolutions)

Agriculture has moved from being

Encouraged to engage in the

Management of

Groundwater quality and quantity,

To now being required to

Actively engage in and being transparent about

Protecting groundwater

- quantity and quality -

as part of

California's integrated water system

For many future generations to come.

There are two overarching themes that

Challenge this revolution of a more

Water regulated agricultural industry:

One.

It is a revolution.

For agriculture - a painfullly new way to work,

For the regulatory agencies - finding that old paradigms don't apply,

For science - a new frontier

There are no ready-made answers and

All parties involved -

Agriculture, environmental stakeholders,

Regulatory and policy decision makers,

And scientists -

Are challenged to find ways to make

This revolution work.

Two.

We must succeed in this revolution.

Irrigated agriculture around the world,

Not unlike that in California,

Provides 40% of all global

Food, feed, fiber, and (bio)fuel.

Finding sustainable ways to

Use and protect groundwater

That we drink and with which we irrigate crops,

If only in drought,

In California and similar regions around the

World is an essential cornerstone to

Sustainable global food security

In and beyond the 21st century.

Thomas Harter

Sierra Nevada

Mountains

Watersheds

Dams

Lake Tahoe

Fires

Recovery

Science

Snowpack

Water Quality

Mono Lake

Forests

Sierra Nevada

After Fire, Restoration

Stephanie Taylor

Vastly scorched areas, sterilized of life-producing plants, must depend on snow, rain, and the wind. And on man.

The Rim fire devastated more than 257,000 acres of the Sierra Nevada and part of Yosemite National Park in August 2013. Because winter rain and snow never arrived, some remote areas of the fire burned throughout the following winter. I wondered about the watershed, the quality of the water and how toxic water affects rivers and reservoirs. I wondered how much water passes over steep mountainsides causing slides, and how much might sink into the ground to regenerate new growth.

I wondered how forests recover from such catastrophic infernos.

One answer can be found in an experimental research forest east of Sonora near Yosemite, in an area that hasn't experienced a fire since 1889.

Eight years ago, Eric Knapp, an ecologist with the U.S. Forest Service, found research maps from 1929 in a dusty cupboard in his office. As if on a treasure hunt, he searched the National Archives and found corresponding records in old leather-bound ledgers. He followed the maps into the Stanislaus-Tuolumne National Forest to the original research sites. He found identification tags still nailed to tree trunks. Knapp remapped each site to compare how the forest had changed over the past 85 years – from soil to shrubs to trees and to the forest canopy.

He wanted to discover how forests best recover from fire and what conditions best create a more resilient forest for the future.

Knapp and his student assistants established new research sites adjacent to the old preserved experimental forest. They performed prescribed burns in the fall. By comparing two main management concepts, they are able to document new growth and wildlife in the two new experimental areas.

Forest recovery a year after fire

In one, they thinned the forest so there was variable spacing, with trees grouped together and open spaces, mimicking how a forest grows naturally. In the second, they cut trees in an even pattern, as is often found where trees have been managed and harvested by man.

One raven calls and several bright orange butterflies provide the only activity in this hushed place.

As we move softly over hot, dusty soil, ash poofs into the air with each step.

Bark of a scorched tree looks and feels like burned popcorn. Ominous holes pit the forest floor, evidence of where mighty cedar, oak, fir and pine trees once stood. I wonder what will happen when it rains.

The ground is deceptive. I'm warned to avoid areas where fire has devoured trunks and followed roots far into the earth, leaving treacherous voids hidden just below the forest floor.

Twenty months after the prescribed burns, I can clearly see the difference between the two experiments. In the first, variable spacing allows sunlight to reach the forest floor. Lilac seeds that can hide in soil for a hundred years have been liberated, sprouting green shoots from barren soil. Green oak leaves sprout from stumps.

In the second experiment, shade from the forest has prevented new growth. I touch the remnants of a pine bough, brittle, sharp and dangerous. The green canopy above that survived the fire provides the only vibrant color in this landscape.

Knapp is optimistic but says, "We don't manage for short-term conclusions, but for long-term. Maybe in five years, we'll have answers."

The Rim fire is a landscape of extremes. Patchy forest and meadows are flourishing with grasses, lupine, lilies and tree seedlings. Vastly scorched areas, sterilized of life-producing plants, must depend on snow, rain, and the wind. And on man.

Or take another 150 years to recover.

Fire at Hetch Hetchy Reservoir threatened water quality for the Bay Area

Mono Lake Lessons

Frances Spivy-Weber
Board Member,
State Water Resources Control Board

I learned that success needs many players, some with essential bit parts; some who must be there every day.

In 1995, I moved from Washington, D.C. to Southern California, and a year later I was the Executive Director of the Mono Lake Committee. I quickly fell in love with the breadth and depth of all things water.

I learned from my work with the Mono Lake Committee, the power of grassroots persistence in the face of larger, better financed adversaries like the Los Angeles Department of Water and Power. I learned that success needs many players, some with essential bit parts; some who must be there every day. The list of legitimate claimants to putting water back into the streams that fed Mono Lake is long, but if we removed anyone who touched the campaign, the outcome would likely have been different.

I learned the value of taking the message of conservation to the homes of Los Angeles communities. When the Mothers of East Los Angeles, Santa Isabel, and other community groups began to go door to door asking neighbors to switch to lower flow toilets, they did their best when they could also show pictures of Mono Lake as one reason to install a new toilet and save water. The community groups knew Mono Lake; they had spent time there with their families. They still do.

Today, I am applying the lessons learned from the Mono Lake experience to climate change and water. Among the challenges the water sector faces from climate change is reduced snowpack, less predictable rainfall, hotter temperatures, more fires and the aftermath of fire, sediment runoff.

What are the Mono Lake Lessons?

• Science is essential to understanding what is happening and is an important guide to planning and achieving desired results in the future. Science, however, is never static, so we must always be listening and learning from our science colleagues young and old. The first scientists to work on the Mono Lake ecosystem were college undergraduates.

Photo: Dave Bundock

Water to protect a special lake or California's way of life is a never-ending challenge, but it is not a challenge just for water wonks.

• The law and courts can be important to setting the stage for future actions. But like science, the law is rarely enough and is not static. Law and the courts change with political variability.

• Communities are at the heart of making societal decisions work. They must be on board, for example, if the state is going to make conservation a California way of life. Every toilet replaced in Los Angeles helped save Mono Lake. Every homeowner and business in California that cuts water used on grass will be the cornerstone of cutting statewide water use, addressing climate change and building resiliency.

• The Mono Lake Committee used messaging help from pictorial magazines like Audubon and National Geographic. Californians have access to apps, Twitter, Facebook, as well as traditional media. Communication is critical.

Water to protect a special lake or California's way of life is a never-ending challenge, but it is not a challenge just for water wonks. The work can (and must) be done by young and old, rich and poor, powerful and those less obviously powerful.

How great is that?

Fran

Sierra Nevada

Lake Tahoe: Fire, Ice and Legends

Stephanie Taylor

A people on land, naturally doing the things that have always been done.

Herman Fillmore, Washoe Tribe of Nevada and California

On a gentle fall day, a simple walk along the shore, through forests of fir and pine ends abruptly. The path expands to a wild, wide meadow and stream. The original Washoe people, the people who have always been here, called it "something that is clear," in their dialect. We call it Taylor Creek.

I try to imagine these people of the Washoe, ghostly figures, not in my memory, but in the memories of others. Preparing seeds and grasses at grinding rocks, fishing with nets and 6-foot funneled baskets for an abundance of Cutthroat salmon, whitefish, clams.

It's as if the tribe has just abandoned this wetland habitat, now empty. It's time to gather pine nuts to the east, and acorns to the west. It's time to prepare winter homes on lower slopes of the Sierra. Maybe that's where they've gone.

There's a cliff within the lake. From a boat floating over the edge, the shore side is bright blue, with rocks and dead trees clearly visible on the sandy bottom. Reflections from the water surface project to the bottom, endlessly undulating.

From the other side of the boat, the cliff plunges relentlessly into dark-blue then blue-black 1,644 foot nothingness. Three major faults lurk. The last major quake was 500 years ago; a magnitude seven occurred 4000 years ago.

Near ice torn valleys on the west shore, volcanoes erupted, lava blocked lake outlets and water began to rise. 50,000 years ago, a monumental landslide occurred under the water and caused an immense tsunami that must have drowned all life around the lake.

Maybe those depths are where the giant, man-eating bird of Washoe legend hides. He's called Ong, and when he flies along the shore, all trees bend with the mighty beat of his wings. Once, Ong caught two men. The second was a clever man who noticed that Ong, while busy eating the other first, closed his eyes as he chewed. Every time Ong opened his mouth, the man threw in arrowheads- all night long while a huge storm raged. By morning Ong was dead. The Washoe say that Ong remains in his nest, hidden deeply within the lake.

If you've ever been swimming in that intensely cold water and looked below, you'd be certain he's there.

Perhaps these depths are where Water Babies hide, sprites that take the form of babies. They are powerful, and can be both dangerous and healing, cause death or good omens, depending on the person. The Washoe bring offerings to these spirits of the lake. The Washoe bring respect.

At the end of the forest, at Taylor Creek, glacial meadows stretch towards high peaks. A drowned tree stands alone on the beach, a solitary sentinel that appears to guard this once abundant place. This place called "something that is clear."

What Lies Beneath: The American River

Stephanie Taylor

For 25 years, I visited this river most days, from flood to drought. On freezing mornings, I followed delicate tracks of tiny creatures on frosty paths, and lost its wild landscape through thick Tule fog. I've inhaled sweet licorice and dusty dryness through sweltering setting suns. I've watched the river rip new channels in winter, then welcome wild iris in spring.

In "The World According to Garp," John Irving refers to the "undertoad," a metaphor for what lies beneath. In the wondrous river that defines Delta watershed history, a dichotomy between the beauty of the surface and hidden forces below is most evident from the middle of the river.

Beautiful, complicated, powerful and deadly, it's like the undertoad. Understanding its forces isn't comforting.

Downstream from Folsom Dam in Sacramento, a hazy warm morning makes for perfect canoeing. In the shelter of Sailor Bar, water is mirror smooth. Our canoe glides silently over weary salmon and cobbles worn smooth. From the bow, the main channel looks wide, daunting, fast and rough. The paddler looks ahead, reading the language of the river. On the surface, ominous raised pillow shapes indicate vertical currents below, while eddies threaten the unwary. Haystacks of angry water indicate hard, shallow objects. The river slaps against the canoe. The paddler looks for a funnel shape in the river and heads for the deepest part of the "V," the safest passage.

Beautiful, complicated, powerful and deadly, understanding its forces isn't comforting.

A river is a living organism, with objectives, memory and behavior. It seeks to balance energy, volume and gradient with the kinds of sediments it carries by regulating width and depth. From the west side of the Sierra, three main channels of the American River feed 1,888 square miles of watershed, flowing first into Folsom Lake and below, to the Sacramento River, and then into the Delta.

From the east, the river responds to the Sierra, from the west, it responds to the tides and the sea, and in between- the Delta.

Folsom Dam was built to manage the conflicting demands of a river that is a victim of the Gold Rush and massive mining operations that filled streams, buried flood plains and moved channels. It was built to control flooding, and for reliable water supplies.

A river is a living organism, with objectives, memory and behavior.

The canoe glides past submerged boulders, cobblestones, trees and exposed roots. Around Sunrise Avenue, towering clay cliffs reveal thousands of years of geology. The river has carved strange mesa structures from clay so hard that it acts like bedrock. Clay defines this river as it cascades through the San Juan Rapids, creating haystacks, eddies, whirlpools and boils. Seeking a balance between erosion and deposits, it cuts into the cliffs on one side, and places gravel on the other. In high volume, the river wants to cut a deeper channel, but the clay resists. When it can't incise, the river widens. Up to the levees.

At William Pond, the river runs around and over tiny islands, spinning in and out of channels with water sparkling in the late afternoon sun. Hard, carved clay bottom and banks lay exposed. Just inches under the canoe, rocks are packed as tightly as paving.

The influence of the sea can often be felt upriver in its ebbs and flows. Near Howe Avenue, the paddler feels the resistance of the rising tide. From the bow, I sense the power and promise of the river.

The undertoad lurks beneath.

The Southland

Drinking Water

Reliability

Colorado River

Mulholland

SWP

Conservation

Tap Water

Bottled Water

Recycled water

Hoover Dam

Imported Water

North vs. South

The Southland

Imagining Paradise

Stephanie Taylor

Who ever brings the water brings the people.

William Mulholland

The Tehachapi Mountains, viewed from the ascending slash of the Golden State Freeway at 70 mph, is hardly my idea of a beautiful landscape. Rising from the agricultural abundance of the Central Valley, these bleak, brown, steep hillsides are a visual shock.

At the top of the Tehachapis, water pumped from the state's California Aqueduct descends into the Southland to meet the urban demands of millions of people. It splits into two branches. Surprisingly, most of the water flows east to dry desert communities, and only about a third makes its way west to Los Angeles and coastal areas south.

From the headwaters of the Sacramento River at Mount Shasta, through the Delta and the Central Valley, I've followed the water to the Southland. And now, the West Branch of the aqueduct has brought me to the San Fernando Valley -- hot, dense and dotted with thousands of swimming pools.

The original Los Angeles River starts here. Water inspired settlement by the original peoples, and with an advantageous climate, has since attracted millions more. Restrained for flood control, it winds through downtown Los Angeles and sometimes flows to the sea.

The Santa Monica Mountains repeat the Tehachapi vistas of drought tolerant plants, a contrast to what lies south in the greater Los Angeles Basin, my home for 18 years. Hollywood Hills, Santa Monica, Beverly Hills, downtown, UCLA and the beach, stunning areas boasting extravagant flowers and gardens, long avenues of trees and lavish lawns, golf courses, parks and ponds.

Fan palms, those ubiquitous icons of paradise, are native plants from ancient tropical times that survived on water seepage from faults.

Fountain honoring William Mulholland, near Griffith Park

Natural growth hints at the future restoration of the Los Angeles River

Near Griffith Park, the William Mulholland Fountain, a memorial to the man who brought water from the Owens Valley to Los Angeles in the early 1900s, symbolizes the eternal desire for water in Southern California. It's a tribute to all the visionaries who imagined this paradise.

A paradise created could easily become a paradise lost.

In the fountain, water from Northern California mixes with the Owens River, the Colorado, storms and wells. Water reaches for the sky, cascades over aqua green tiles and recycles again. It represents a powerful and contentious past- and a vulnerable future dependent on reclamation and conservation, and threatened with unpredictability. It represents adaptation, innovation and collaboration.

The Los Angeles River is just across the street. Its concrete channel is being revitalized with habitat and integrated into the recreational vitality of the region.

The East Branch of the Aqueduct delivers water to thousands of spreading developments, to the Inland Empire, Palmdale, San Bernardino and Riverside, where communities must meet the challenges of the built-in thirst of desert heat. Like the west side of Los Angeles, water districts are incentivizing conservation with rebates and other programs such as reimbursements for drought tolerant landscape, mulch and drip.

For five months and 700 miles, from the headwaters of the Sacramento to the Los Angeles River, I've followed this complex, miraculous water system. Its aging infrastructure is supported by tens of thousands of people charged with making it efficient and safe from flooding, contamination and failures.

Mulholland said, "Who ever brings the water brings the people." But even he understood limits.

Every Californian must prioritize water as a critical resource. What was a paradise created could easily become a paradise lost.

Children in the Los Angeles River channel

Southern California: The Search for Water

Rita Schmidt Sudman

Today, many new sources of water for Southern California involve programs rather than delivery of water by concrete.

Most Southern Californians have no idea that much of their water travels hundreds of miles to reach their taps. Long-time regional rivalries over water still exist. So deeply held is the belief that people living south of the Tehachapis waste water for swimming pools and lawns, that many Northern Californians have trouble believing the fact that urban Southern Californians regularly use much less water than they use.

Since the 1982 defeat of a proposed canal around the Delta to improve their water supply, Southland water managers have found new programs, including conservation, to be a major way to get "new" water. It is a relatively new phenomenon that some of the ways to increase water supply emphasize the soft path of incentive programs, rather than traditional dams and canals. Throughout most of the history of Southern California, the plan was to import water to support growth.

Importing water begins

There was no need to import water before the Americans arrived. Los Angeles was a sleepy Mexican village when gold was discovered, while San Francisco and Sacramento boomed. When statehood was achieved in 1850, only 1,610 people lived in Los Angeles and water wasn't much of a concern. In 1878 when William Mulholland arrived, Los Angeles' water supply came from an open ditch below where the laundry was beaten. He promptly made major improvements.

It was not long before lack of water became the major problem in Southern California. In the Spanish and Mexican periods, drought often crippled the Missions, almost bringing starvation at times. Water from the coastal mountains was not dependable. Fortunately, there were natural artesian springs in many areas; hence town names like Artesia and Spring Valley.

The state's California Aqueduct brings water to Southern California.

In 1870, much of Southern California remained in enormous land holdings dating from the Rancho era. Grazing was the main economic activity, but the two droughts between 1856 and 1864 finished off most of the cattle ranching.

In 1877, the Southern Pacific Railroad connected Los Angeles to the East. The Ranchos started breaking up by the time settlers arrived on the Santa Fe Railroad, in 1885. Citrus was the first lure. Since springs were drying up and the watershed streams were unreliable, deeper wells were dug. Early in the new century, it was becoming evident that there was a severe overdraft problem in many areas – farmers were using more water than nature replaced, and thus the water table declined.

Water from the Owens Valley

By the time Mulholland became Los Angeles' chief engineer, he was looking for a new water source for the growing population. It is often said that it was not movies, aircraft, agriculture and other industries that invented the Southern California we know today; it was Mulholland who should get the credit because he brought the water. He found the water in the Owens Valley, 250 miles north in the Eastern Sierra. After voters overwhelmingly approved a bond, an aqueduct was completed in 1913, bringing water to the San Fernando Valley and Los Angeles. Los Angeles began to catch up in population with San Francisco.

Importing water is not just a Southern California phenomenon.

Importing water was not just a Southern California phenomenon. In 1913, the same year Los Angeles began importing water from the Owens Valley, Congress authorized San Francisco to build the aqueduct from Yosemite's Hetch Hetchy Valley.

With the building of the Los Angeles Aqueduct from the Owens Valley, Los Angeles' water supply problem was expected to be solved for many years. Planners were shocked when they learned population growth would make the system inadequate within 10 to 15 years. Since Owens Valley water wouldn't be enough to quench the Southland's thirst, the search was on for another source. Mulholland was part of that search when he travelled to the Colorado River. In 1924, Los Angeles applied to divert 1.1 million acre-feet annually from the Colorado River.

The Los Angeles Aqueduct brings water from the Owens Valley.

In fact, the Metropolitan Water District of Southern California (MWD) was created in 1928, to act as the water wholesaler for the Southland. Eleven cities, including Los Angeles, formed the agency to build the aqueduct. Currently, MWD has 26 member agencies.

Water from the Colorado River

Importing Colorado River water would help stop declining groundwater levels. Voters in 1931 passed a $220 million bond issue to build the 242-mile aqueduct to coastal areas. It was completed in 1941, in time to meet the water needs of burgeoning defense plants, and the people relocating to work in them. The water and power from the Colorado River's newly built Hoover Dam would help win World War II. To ensure water for the Navy in San Diego, President Roosevelt caused what some historians have called "a shotgun marriage" between MWD and San Diego. Eventually water was imported and distributed from Ventura to San Diego.

Water from the Colorado River also was thought to ensure generations an adequate supply. However, the boom years did not end with the war. Too many people from other parts of the country got a taste of Southern California, like my parents from cold Wisconsin, who liked the sunshine and oranges. My Marine Corps dad was stationed in San Diego before and after WWII. More water was needed for the new transplants and their growing families like mine.

In 1958, Pat Brown said, "What are we to do? Build barriers around California and say nobody else can come in because we don't have enough water to go around?"

Enter a Northern California politician who would become a new Mulholland and help bring more water to Southern California.

When Pat Brown became governor of California in 1958, 1,000 people a day were coming into the state and straining resources – 80% of those settling in the job-rich Southland. In 1958, Brown said, "What are we to do? Build barriers around California and say nobody else can come in because we don't have enough water to go around?"

Water from the State Water Project

His first priority as governor in 1959, was the passage of the State Water Project (SWP). He used his political skill to get the proposal on the 1960 ballot, and convinced voters to ride the New Frontier electing John F. Kennedy President and ratifying a $1.75 billion bond issue to finance the massive project. Although the vote for the project was four to one favorable in the south, the narrow statewide vote passage showed that strong regional – North vs. South – animosity existed. Water now would be brought from the north and lifted 2,000 feet over mountains , higher than water is lifted anywhere in the world. By the 1970s, the 444-mile aqueduct was bringing water from the Sacramento Valley through the California Delta into an aqueduct, named for Pat Brown, and over the Tehachapi Mountains into Southern California.

After the 1977 drought, Southern California interests accelerated their efforts to bring more water south. By 1980, the south depended on water from the north for roughly half of their water supply. A proposal to getting more water south did not get enough support. In 1982, in a referendum vote on a Peripheral Canal, the proposal to carry water around the California Delta and on to Central and Southern California, was defeated. It was the first time voters rejected a plan to bring water to Southern California. It passed in Southern California, but with lukewarm support, and failed in most of the rest of the state with votes up to 90 percent opposed.

Jerry Brown was then governor and his philosophy was not the same as his dad's. He supported the proposal to build the canal around the Delta in 1982, but not, as I remember many water leaders saying at the time, very enthusiastically. His was a more "small is beautiful" philosophy that recognized natural resource limitations. Also the environmental movement, nonexistent in his father's era, was gaining in popularity and political power. Perhaps recognizing the changing times, the father, who I got to know at that time, was careful not to criticize his son.

The Colorado River Aqueduct brings water from the Colorado River to Southern California.

State Water Project water is pumped higher than any place in the world, more than 2,000 feet over the Tehachapi Mountains into Southern California.

Water for the 21st Century

So the winds shifted. The idea of less dependency on water from the North gained support. After the 1982 vote, MWD did not go beyond their area in search of the next watering hole. MWD and Los Angeles stepped up incentive programs through toilet rebates and other conservation programs, established water market transfer programs with farmers on the Colorado River, and found other ways to diversify their water portfolio including MWD building a major reservoir, Diamond Valley Lake, to store water in their own backyard.

Many Southland water agencies began to work together to integrate their water management. Conservation has been a key part of the new programs. Now serving 19 million people, MWD is spending $450 million for the nation's largest turf removal and water conservation program. Over the next decade this is expected to generate enough water savings—by removing 175 million square feet of lawn - to nearly fill another Diamond Valley Lake.

San Diego County Water Authority also diversified its supply. In the early 1990s, a drought reinforced the fact to San Diegans that they were at the end of all the pipelines. In 1991 the Water Authority bought 95 percent of its water from MWD. With diversification and a water purchase of conserved Colorado River water from the Imperial Irrigation District, the imported water eventually will be about 50 percent. The Carlsbad Desalination plant, online in 2016, will supply about 7-10% of San Diego County's water. The desalted water is expensive, at about $2,000 per acre-foot, but reliable. The cost gap could diminish as technology improves.

In spite of these programs, Southern California still seeks to bring in a significant amount of imported water in the coming decades because conservation and the other programs will not be enough. MWD looks to ensure a more reliable imported supply from the SWP by paying for a controversial program to pour concrete and build tunnels under the Delta to send water directly to the state aqueduct.

Forty years after his first term as governor, Jerry Brown supports this tunnels project to continue the long-ago legal commitments of his father and voters to bring water into Southern California. In spite of philosophical differences between father and son, Jerry Brown has said that if he'd been governor in that earlier era, he would have done the same things his father did. He notes, "Who sets the agenda? The times set the agenda."

Jerry Brown said, "Who sets the agenda? The times set the agenda."

I think both Browns were governors who responded to the needs of their time. Both governors recognized that some water always will flow from north to south. But today, much of the new sources of water for Southern California involves incentive programs other than delivery by concrete. One certainty is that regional fights over water will always be with us.

As Pat Brown once told me, "Rita, water with problems is better than no water at all."

Doused by Water

Maureen A. Stapleton
General Manager,
San Diego County Water Authority

As I found out over the next eight years of negotiations, monopolies don't surrender power, money or control easily.

I was on a career path to become city manager of San Diego when I was doused by water.

During my 20 years in municipal government, including 10 years as assistant city manager for the City of San Diego, I never had water utilities within my span of authority. So, it was quite a surprise when I was contacted in late 1995, by a recruiter charged with finding the next general manager of the San Diego County Water Authority. The longtime manager of the Water Authority, Lester Snow, had left the agency months before to head up the CALFED's Bay-Delta Program. John Lockwood, the recently retired San Diego city manager, had been brought in by the Water Authority as interim General Manager.

John and I had worked together at the city, and knew each other well. He mentioned that the Water Authority had announced, weeks earlier, a proposed agriculture-to-urban water transfer between the Imperial Irrigation District and the Water Authority. I had negotiated many major development agreements at the city, and John predicted I would lead and wrap up negotiations on the deal in a matter of a few months. That sentiment was affirmed by Water Authority Board Chair Mark Watton, who assured me that "this little transfer deal" would go smoothly and was not much to worry about.

I was duped.

Coming from municipal government, I thought public water agencies would be very similar in structure, management and operational characteristics, making the transition easy. I was wrong. At that time, water was steeped in tradition, insular, heavy on engineering for solutions, often isolated from its ratepayers, monopolistic, and rarely covered by the media. Lawsuits lasting decades or more were not uncommon. "Water's for fightin" was, I discovered, more than just a shopworn cliché.

As it turned out, that "little transfer deal" was the spark that not only ignited landmark changes in Western water, but also a fierce water war that challenged the power and authority of a number of water agencies. As I found out over the next eight years of negotiations, monopolies don't surrender power, money or control easily.

I've been a part of an industry transitioning from command and control to partnerships and collaboration.

But, what emerged in October 2003 as the landmark Colorado River Quantification Settlement Agreement changed – for the better – how California's Colorado River resources are managed and protected. Its reach extended beyond Southern California's Colorado River water agencies to include groundbreaking accords among the seven Colorado River basin states, and between the United States and the Republic of Mexico.

It's been 20 years since I began my career in water. Over the past two decades, I've been a part of an industry transitioning from command and control to partnerships and collaboration. I remain hopeful about the future of water in California, especially for those agencies and regions that embrace change, implement bold new ideas and strategies, engage their communities, and do the hard work of being a leader in an industry that continues to challenge us every day.

San Diego

Drinking Water – Tap versus Bottled

Rita Schmidt Sudman

Bottled water is a packaged 'food,' regulated under the FDA, while tap water is regulated under the EPA, with legally enforceable standards.

Drinking lots of water is key to good health. Water is often called the river of nutrients to our cells. Every year Americans choose to drink more water—bottled water. We drink 32 gallons of bottled water each year, up from 21 gallons in 2003. After soft drinks, bottled water is now the second most popular drink in America. It outsells milk, beer and coffee. According to one industry group, it could soon be the number one drink.

There are differences between tap and bottled water due to regulations by two different federal agencies. Bottled water is defined as a packaged "food," and is regulated under the federal Food and Drug Administration (FDA). Tap water is regulated under the federal Environmental Protection Agency (EPA) which has legally enforceable standards that apply to public water systems. In 2009, the federal General Accountability Office released a report saying bottled water manufacturers are not required to disclose as much information as the public water agencies, pointing out that there are gaps in federal oversight authority.

People often ask me whether they should drink tap water or bottled water. Personally, I drink tap water in the United States because I trust it. Sometimes I drink bottled water for convenience when traveling.

Here are a few facts you might consider in making your choice.

Bottled water manufacturers are required to maintain testing records to show government inspectors that their water can pass tests used by the FDA in its own laboratories. As with other foods, the FDA periodically collects and analyzes samples of bottled water, especially if there is any previous history of contamination, consumer complaints or if the product originates in an another country. The FDA says, since the bottled water industry generally has good safety

Unlike tap water in some countries, Americans can have confidence in the integrity of the public drinking water systems in their homes.

record, manufacturers have been assigned "a relatively low priority for inspecting."

Unlike tap water in some countries, Americans can have confidence in the integrity of the public drinking water systems. Tap water from public systems is regularly treated and disinfected to remove particles, chemicals, bacteria and to meet standards. The EPA mandates that public water providers give residents detailed accounts of what's in their tap water and the results of regular testing. These are called customer confidence reports. You may have seen these detailed reports as an insert in your home water bill. These report the source of your water, evidence of any contamination and compliance with regulations. This accountability holds tap water to high scrutiny. Bottled water companies are under no federal obligation to deliver such information to the consumer—on the bottle or in any other way.

40% of California schools don't offer free drinking water in lunch areas.

Where does bottled water come from?

About 75% of the water comes from groundwater wells. Further refined tap water makes up the remaining 25%. Although the bottled water produced in California's 110 bottling plants is only .02 percent of the water used in California, it has gotten attention during California's latest drought because its production uses local water sources. Out of public concern for water being taken from groundwater in California, Starbucks announced in 2015 it would move its bottled water production from California to Pennsylvania. Another issue to consider is the effect on the environment of millions of tons of plastic used in bottling billions of bottles of water each year.

Lead contamination

In addition to the debate over tap versus bottled water, people drinking from private wells should consider other factors. If you have well water that doesn't go through a public system, there could be contamination. Your private well should be tested every year. Plumbing also can affect water delivered to our homes. Old lead pipes can leach lead into water. If you think this is an issue in your home, you should have your water tested, though lead service pipes to homes are not common in California. The state has tested extensively for lead in drinking water since 1991. Call the State Water Board to get certified lab numbers.

The crisis in Flint, Michigan was caused by the change to an inferior water supply with a high pH that caused corrosion in old lead pipes. Flint failed to implement appropriate corrosion control measures required by federal law.

A need for free, clean water from public fountains

Public drinking fountains go back at least to Roman times. The modern American fountain was created about 100 years ago in Ohio by Halsey Taylor whose father died from Typhoid Fever, a waterborne disease. Taylor invented a double bubbler drinking fountain that was more sanitary. It used two separate water streams converging at a higher angle to create a pyramid of water.

Today, many public drinking fountains are in disrepair, perhaps due to people drinking more soda and bottled water. There is less demand to maintain clean, safe fountains. This is especially a problem in our schools. Even mild dehydration can affect student learning, mental alertness and physical performance. It's clear that school children need to drink more water. For free. Schools get revenue from soda vending machines and reimbursements for milk under federal programs. There are no incentives to provide free drinking water.

There are no incentives in our schools to provide free drinking water.

Because of the poor conditions of many school drinking fountains, in 2011, a state law was passed requiring California schools to provide free drinking water in cafeterias. Surveys show that 40 percent of schools don't offer students free drinking water in lunch areas. Consequently, only 15% of middle school children drink an adequate amount of water, according to the Centers for Disease Control that says children should be consuming at least half their total water intake while in school. Also, a federal law now requires schools to make water available to all children who are part of the National School Lunch Program. Although these laws intend to encourage school districts to provide water, there are fairly easy ways for California schools to opt out, particularly from the state law.

However you choose to drink water, you're entitled to trust its quality.

Why don't children just bring water to school in refillable bottles? The reason often is that many schools don't allow children to bring liquids into their schools, fearing something harmful may be in the container. Hopefully, those rules are changing.

You may have seen the new tap water bottle filling stations at airports and other public locations. These stations provide clean and convenient drinking water. In San Francisco, a partnership between the water utility and the local school district provides opportunities for people and students to fill their own bottles at several parks and schools. The water utility also includes signage to help students connect them with the source of their water and to learn about their watershed.

However you choose to drink water, you're entitled to trust its quality.

Every Generation, a Southland Water Milestone

Jeffrey Kightlinger
General Manager, Metropolitan Water District of Southern California

The State Water Project needs a retrofit and reinvestment in the Delta so that water can continue to be pumped and conveyed in greater harmony with fish species.

Driven by growth and the eternal draw of California, Southern California has needed to write a new chapter in its water history every generation. The next chapter is now coming due. The elements of the plot are emerging, from the historic effort to solve the water system/ecosystem crisis in the Sacramento-San Joaquin Delta to bold proposals to increase local supplies and lower demand. The region is moving forward based on where it has been, with lessons learned along the way.

At the turn of the 20th Century, all of California's fast-growing urban centers – San Francisco, the East Bay and the city of Los Angeles – sought to import supplies from the distant Sierra Nevada mountain range through brilliantly designed conveyance systems powered solely by gravity. San Francisco tapped the Tuolumne River. The East Bay looked to the Mokelumne. Los Angeles secured its supply from the Owens River on the Eastern Sierra via its 233-mile aqueduct system, designed by William Mulholland. Its completion in 1913 began Southern California's importation of water.

The subsequent generation of stunning growth, far beyond Mulholland's original projections, set the stage for the birth of Metropolitan. A second generation of imported supply – this time Mulholland eyed the Colorado River – would soon be needed, but Los Angeles had neither the financial capacity nor the full need for such a project on its own. Instead, the urbanizing areas of Los Angeles and Orange counties merged forces through a new regional water agency, the Metropolitan Water District of Southern California. The California Legislature created Metropolitan in 1928. Southland voters in the throes of the Great Depression approved $220 million in bonds to build the 240-mile Colorado River Aqueduct.

The final installment in Southern California's expansion of its imported supply portfolio came another generation later in 1960. Governor Pat Brown proposed the State Water Project, construction of the world's tallest earthen dam on the Feather River, and the California

Photo: CA Department of Water Resources

Aqueduct to transport supplies from the Delta to the foot of the Tehachapis. Metropolitan partnered with Governor Brown, providing the financial cornerstone of the project by guaranteeing 50% of the project's cost for 75 years. California voters narrowly approved the project in 1960, a sign of the political divisions that were to come.

> *I grew up in Southern California and went to Berkeley. I was flabbergasted [to find] it was 'hate LA.' It's not as virulent today as it was. Northern California is relying on more imported water as well, so they've realized we're in this together.*
>
> *From an interview in the Los Angeles Times, April 2015*

A generation later, an epic drought transformed Southern California water in 1990. The assumption that the region's elaborate and diverse water importing systems would always provide sufficient water was debunked. Metropolitan reinvented itself, establishing conservation as an official pillar of its long-term water portfolio and creating a network of reservoirs and groundwater banks to capture supplies in wet years. Today Metropolitan has 14 times the overall storage capacity than it did in 1990, more than Lake Shasta. The region now imports less water than it did during the peak of consumption in 1990 despite growing by 5 million people. Funding development of local supplies such as water recycling and reclamation throughout Southern California was built into the cost of imported water. Every gallon of water to meet demand from population growth now comes from lowering demand and increasing local supplies.

Colorado River Aqueduct delivering water to Southern California

Once again, we find ourselves reaching the end of the last generation's planning effort

This new portfolio approach to water supply - balancing imported water, local supply development and demand management - has served the region well these past 25 years. Southern California has continued to enjoy spectacular economic growth and diversity with assured water security. Regional growth has been managed on a water demand neutral path. But once again, we find ourselves reaching the end of the last generation's planning effort.

The Metropolitan water portfolio is expanding and diversifying.

Southern California now faces a familiar, historic crossroads in water planning. The State Water Project needs to modernize and reinvest in the Delta so these vital supplies of high quality water can continue to be pumped and conveyed in greater harmony with fish species. Desalination, recycling and groundwater cleanup projects are also emerging local priorities. The water portfolio is expanding and diversifying. Some see an "either/or" choice between one water investment versus another. Others see an "all of the above" as the only reliable path. It will be the careful and clear prioritization of all these options into a comprehensive water supply road map that will be the key to success for the next generation of Southern Californians.

Xeriscape landscaping is creating "new" water for Southern California.

One Water One Watershed: The Answer for 21st Century Water

Celeste Cantú
General Manager,
Santa Ana Watershed Project Authority

Many say water in California is complex. Compared to people, water is easy.

Drought, not unusual for California's Mediterranean climate, but made extreme with climate change affecting much warmer temperatures, jumpstarts evaporation causing plants and people to need more water. The Delta, through which water is moved from the North to the South, is near collapse and the Colorado River is overdrafted and will go into "shortage" soon-triggering events agreed to years ago when no one ever thought shortage would happen.

As we continued to grow, water's Four Horsemen of the Apocalypses arrived in California.

None of this is new to California. Growing up hearing tales of early Californians, ranchers, my relatives were driven from California as their cattle died during severe drought in the 1860s. Nineteenth century water management in California is best described by this painting – locate your ranch or later town around water and protect it from everyone else.

20th century water management built large pipes and pumps to move water from where it was to where it was needed to fuel a new economy. I remember my father, grandfather and uncle debating the salinity content of Colorado water deliveries to Baja California. I dismissed water as the realm of old men only to be snagged later. I was right. Unlike the rest of the world, old men "Water Buffalos" controlled water in California. But this is changing.

The 21 century brought challenges to water management, quality, scarcity, environmental degradation and cost. This perfect storm demands innovation, cooperation and collaboration. The Santa Ana Watershed Project Authority (SAWPA), created in 1968 to keep peace among then warring water districts and located in the counties of San Bernardino, Riverside and most of Orange County, took a different approach. SAWPA looked to integrate the different water uses with a goal that one use would not impair the next use and that those same water drops would be used repeatedly as they flowed from the headwaters to the ocean.

19th Century California Water Management

Fight for the Waterhole by Frederic Remington

21th Century California Water Management Strategy

Water is used many times by families, fish and farms as it flows from headwaters to the ocean.

Ratepayers were tired of paying flood control districts to capture stormwater and quickly channel it out of their community to protect people and property, while paying a premium to their water district to find water hundreds of miles away and transporting it at great carbon footprint into their communities while paying their wastewater district to highly treat their waste stream and dumping it into the river.

They said, "Get your act together." We are willing to pay only once.

SAWPA created One Water One Watershed (OWOW) in 2007 to create opportunities where different water uses can be harmonized, synergies and efficiencies created. If we were to consider all the budgets in the aggregate and make investments strategically to cascade benefits throughout the watershed, we would find efficiencies.

An investment made high in the watershed can pay dividends cascading down to serve the entire watershed. Rather than hardscaping waterways to move flood water to the sea, slow it down to percolate to replenish aquifers. We had science and technology to separate the H2O molecule from what contaminates we put into it.

We learned to speed up Mother Nature's cleaning process and manage water for people and wildlife more effectively.

OWOW integrates water management horizontally across political boundaries, coordinating across city and districts limits and vertically by integrating different disciplines, which sort water into different realms such as flood, water supply and wastewater. These old divisions no longer serve our purposes today as we look to storm water and highly treated wastewater as drinking water sources.

Together we have achieved new levels of cooperation such as managing water like a commons for multiple benefits.

- We agreed to pool funds from throughout the watershed to store groundwater during wet times in the upper watershed to be called on by all in dry times.
- Forest First, a MOU between the US Forest Service and water districts to manage the forest, where 90% of our precipitation falls, specifically to improve water supply and quality.
- Thinking of water as a shared resource as opposed to water as gas to be extracted, delivered, consumed and reduced to energy and greenhouse gases. Water is not consumed but returns to recycle again.

The water drop circulates from atmosphere to rain and snow, melting into tributaries and percolating into the aquifers and rises again in to the rivers that define the watershed to sustain life, aquatic, human, wildlife and crops before it reaches the ocean where it begins the never ending process again. The promise of Integrated Water Resource Management is that we can manage to keep it clean, and increase the beneficial uses as it cascades from the head

One Water One Watershed is an integrated planning process within the Santa Ana River watershed.

waters in the mountains.

It's been said the water drop in the Santa Ana River uniting San Bernardino, Riverside and Orange Counties is the hardest working drop in California as it serves six million Californians one by one and each time cleaned up and returned back to the River to be used again by fish, and the next community before it reaches the sea.

It's been said that the water drop in the Santa Ana River uniting San Bernardino, Riverside and Orange Counties is the hardest working drop in California.

This 21st century thinking can be replicated throughout the West. The stakes are high and the obstacles are egos, turf, and control. It requires people to suspend opportunistic ways, need to control and to share governance to create synergies and efficiencies.

Many say water in California is complex. Compared to people, water is easy.

Celeste

Imperial Valley & the Salton Sea

Mexico

Water Transfer

Colorado River

Salinity

All-American Canal

Desert

Birds

Public ownership

Irrigation

Salton Sea

Love Story

Farms

Great Flood

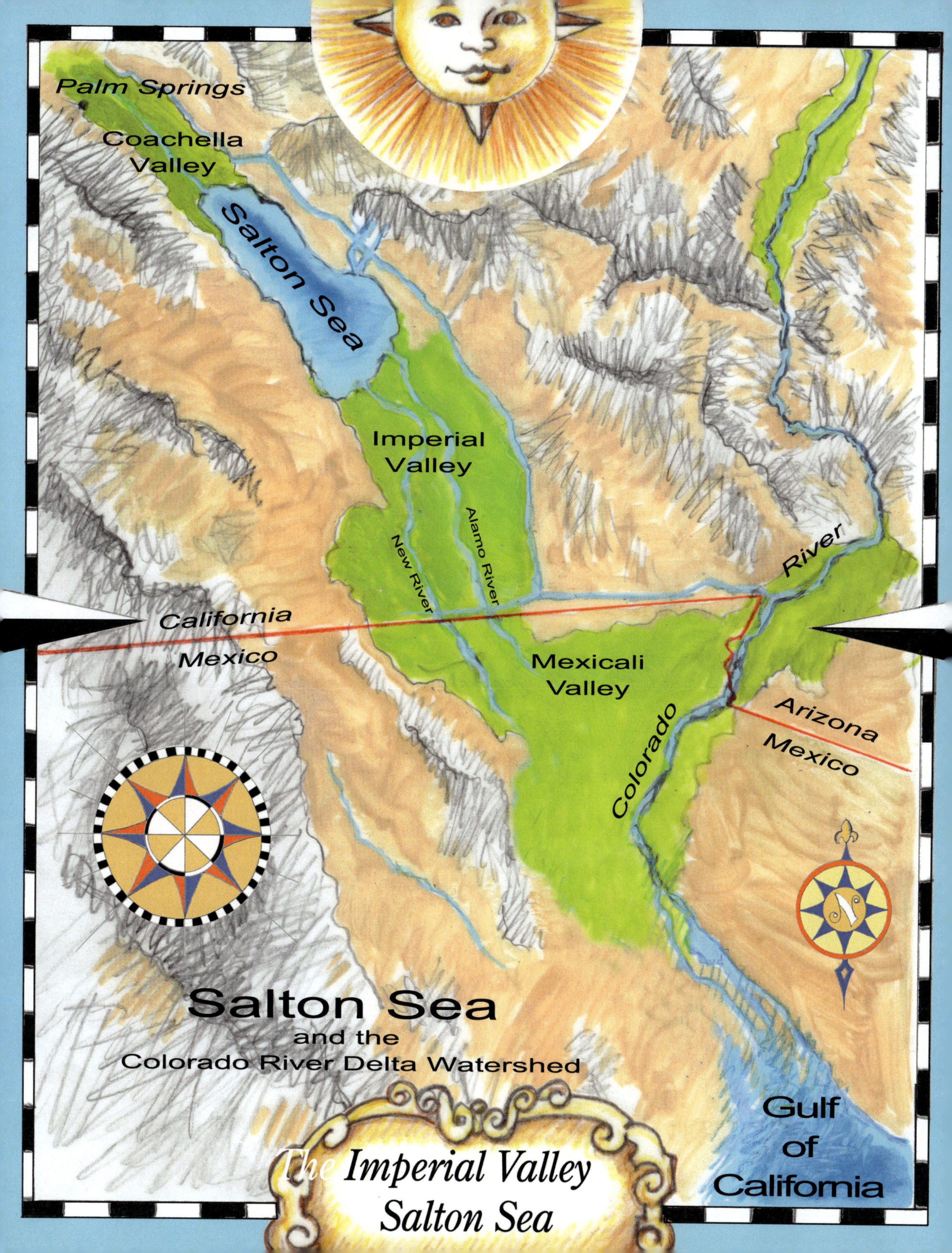
Palm Springs
Coachella Valley
Salton Sea
Imperial Valley
New River
Alamo River
California
Mexico
Mexicali Valley
Colorado
River
Arizona
Mexico
Salton Sea
and the
Colorado River Delta Watershed
Gulf of California
The Imperial Valley
Salton Sea

Imperial Valley & the Salton Sea

Vanishing Dreams

Stephanie Taylor

Three flows are important for the Salton Sea: water, wildfowl and money.

Jay Lund, UC Davis Center for Watershed Sciences

The North Shore Beach Yacht Club opened in the early 1960s, attracting a glamorous crowd – Frank Sinatra, Jerry Lewis, the Beach Boys and many others. An hour southeast of Palm Springs and a few hours drive from Los Angeles, California's largest lake attracted thousands of visitors who came to fish, camp and boat. Salt water provided a perfectly bouyant venue for festivities and regattas.

Today, the boats are gone and so are the stars. Choppy, steel-colored water rolls up beaches bleached with salt. Gulls hover, one white egret hunts.

It appears to be a forsaken landscape.

But there is hope for the rehabilitation of the Salton Sea and surrounding environs. The yacht club, flooded and abandoned since the 1980s, was renovated in 2010 as a community center. The maritime themed building boasts fresh exteriors in yellow, aqua, nautical blue. Flags signal bright spots of optimism, as experts race against time to save this desert lake and avoid environmental disaster.

Long ago, locals called this place "the valley of death." At the turn of the last century, visionaries diverted the Colorado River, and transformed the south and north ends of the basin into the Imperial and Coachella valleys. Developers brought settlers who irrigated hundreds of thousands of acres, turning desert into productive farmland with long growing seasons, especially for winter fruits and vegetables.

Sustained by little rain, a few rivers and agricultural runoff, the Salton Sea has been out of balance since the '80s. A fertilized soup of selenium, nitrate, phosphate, chromium and algae blooms, the lake creates toxic conditions for birds, fish and humans. Evaporation and water transfers intensify concentrations of salt.

A friend of mine visited the lake years ago. She described "the unbelievable stench … Dead fish lined the shoreline, and heading into shore were the flopping bodies of fish in their final throes. And farther out, you could see fish jumping, as if they were trying to get the hell out of that water."

On a cool day in spring, the Salton Sea conceals the drama of summer heat, dust, die-offs, odor and crisis. Neighboring wetlands, one of several lush habitats around the lake, shelters wildlife. Year round, more than 400 bird species navigate the Pacific Flyway from the Arctic to Argentina. Millions of birds stop here, floating, flying, feeding and nesting.

The Salton Sea has been a place more notorious for beauty forsaken than for opportunities.

Jay Lund, of the University of California at Davis Center for Watershed Sciences, says that "three flows are important for the Salton Sea: water, wildfowl and money." Cleaner water, in the form of treatment facilities, salt evaporation ponds and wetland restoration, will benefit waterfowl.

As for money, any strategy will cost a lot.

The rehabilitated yacht club is a symbolic step toward fixing the Salton Sea. No one solution, but a concert of collaboration may foster equilibrium. Imagine what this place could be, full of life and possibility.

A Desert Becomes the Imperial Valley and the Salton Sea is Created

Rita Schmidt Sudman

The next time you drive between Los Angeles and Phoenix, think about the epic story of this arid land that's so important to the nation.

Irrigating the desert is a dream as old as the Bible. To turn the Colorado Desert into the productive Imperial Valley took a great transformation of land and water. The next time you drive between Los Angeles and Phoenix, think about the epic story of this arid land that is so important to the nation.

The opportunity to convert the desert into farmland was seen early in California history. Before the Civil War, there were plans to build a gravity canal from the Colorado River. The idea was that the land around the Salton basin, site of an ancient dry lakebed, could be irrigated with the Salton sink used for drainage. At this time there was no Salton Sea. In 1892, Engineer C.R. Rockwood was hired by a private land development company to survey a proposed project. Financial issues dogged this and other irrigation plans.

Irrigating the Colorado Desert

It took the genius of George Chaffee, a Canadian engineer who had successfully developed the Southern California irrigation communities of Rancho Cucamonga and Ontario, to bring the Colorado Desert irrigation plan to life. The bronze relief wall at the Ontario airport pays tribute to him. He was successful in establishing a development company in which lot buyers were given proportional shares and got electricity from the canal's hydroelectric generator. By 1886, the Chaffey development became the gold standard for private water companies.

After an unprofitable irrigation adventure in Australia, George Chaffee returned to California and joined the California Development Company as chief engineer with the idea of irrigating the Colorado Desert. He rightly figured the Colorado Desert was not an attractive name, and so he promptly renamed the land "Imperial Valley." Based on his reputation, people followed him into the desert. Chaffee adopted Rockwood's design to build a canal from the Colorado

Photo: CA Department of Water Resources

Imperial Valley Peppers

River and construction began in 1901. The canal was built partly in Mexico following an ancient streambed that flowed naturally toward the Salton Sink, and back into California.

In four years, the Valley's population went from a handful of settlers to 12,000, and irrigated acreage went from 1,500 to 67,000.

The federal government, represented in the Valley by the new Bureau of Reclamation, was wary of private irrigation companies, and told the settlers that these entrepreneurs could go broke, especially in an emergency situation.

An emergency was on the way. Silting caused continuous intake problems at the canal head. But the real problem happened when floods in 1905 hit the area. For a number of reasons—still debated by historians—involving Mexican and U.S. authorities, Rockwood, now with the California Development Company, made the decision to build a 3,300 foot cut through sandy soil in Mexico without building a control gate. It was believed safe because in the last 27 years the river had only three minor floods.

Soon, five big floods hit. The mighty Colorado River jumped its banks and flowed into the Alamo River, thus flooding the Imperial Valley. Attempts to control the flow failed. The Colorado River turned 180 degrees and rushed into the Salton Sink forming the modern Salton Sea.

The modern Salton Sea is born by accident

The Bureau had gotten it right. The private company did not have the resources to save the Valley. The federal government now rejected assistance as the cut, which caused the flooding, was in Mexico. So it fell to the area's largest landowner, Southern Pacific Railroad, whose tracks also were being flooded, to take drastic action.

It took 52 days of continuous effort to save the Valley. Railroad president E.H. Harriman diverted trains and crews all over the West, and threw 6,000 carloads and tons of rock and gravel into the gap, finally closing it in February 1907, and sending the river back towards the Gulf of Mexico.

The Salton Sink was now the Salton Sea. It would become a defining part of the Imperial Valley landscape and would play important roles in the Valley's future.

The Salton Sink became the Salton Sea, a defining feature of the Valley landscape, and would play important roles in the Valley's future.

As the flood receded, the settlers took control and voted in 1911 to form the Imperial Irrigation District (IID). It was a vote for public rather than private ownership. The IID soon acquired the rights to the bankrupt company. The new district immediately began lobbying Congress for a canal built by the federal government. The 1912 Mexican Revolution helped cement support for a canal built entirely in the United States.

The epic fight had caught the attention of the nation. A popular book and 1926 movie, *The Winning of Barbara Worth* by Harold Bell Wright, was filmed in the Imperial Valley, and tells the story of building the irrigation system, the great flood and a love triangle between an engineer who competes with a local cowboy for the love of a feisty girl. She chooses the engineer who saved the Valley. The story reveals a lot about the prevailing belief that technology could conquer all and improve peoples' lives.

After Hoover Dam was completed in 1935, the Bureau finished the aptly named All-American Canal in 1942, and the Imperial Valley was saved from future disasters. Today, the Valley's gross agriculture production exceeds $1 billion each year. While about half of the Valley's 500,000 acres are in lower-value hay crops, the Valley also produces two-thirds of the nation's lettuce, carrots, broccoli, spinach, onions and other winter vegetables.

Imperial County ranks among the top 11 agricultural counties in the nation.

The Valley produces two-thirds of the nation's winter vegetables, and ranks among the nation's top 11 ag counties.

The Salton Sea has the state's largest lake surface at 375 square miles, but is fairly shallow and 50 percent saltier than the Pacific Ocean. Many birds find the Salton Sea to be a welcome stop on the Pacific Flyway. As farming grew in the Valley, agricultural drainage continued to enlarge the Sea, and sport fishing was introduced, turning the Sea into a popular recreational spot. In the early 1960s, the area was being developed as a recreation spot for middle-class families and even drew headliners like Frank Sinatra to its shores.

However, by the 1980s, increased farm drainage was causing local flooding. Salinity and pollution intensified, and the development dream began to die. Large die-offs of birds occurred in the 1990s. In 1997, congressman and former entertainer Sonny Bono, who had water skied in the Sea's heyday, hoped to restore the Sea. He got a congressional task force formed before his untimely death. His widow, Mary Bono, was elected to his seat and continued the fight but with little success.

An emerging plan to transfer some of the Valley's water would bring attention to the fate of the Sea.

Water marketing brings change

Through the years, the IID secured water rights to a tremendous amount of water. Imperial Valley gets nearly 20 percent of the Colorado River's water. Agreements and court orders had reduced the district's water apportionment at amounts between

Lining the All-American Canal for water conservation transfer

New pepper planting in the Imperial Valley

3.85 and 2.6 million acre feet annually. Before the early 2000s, IID had sometimes used upwards of 3.4 million acre-feet annually. However, that water had never been "quantified" for legal purposes. Whatever the IID's annual water use was in those years, it was a lot more water compared to the much smaller amount of 550,000 acre-feet going to populous coastal cities annually.

California often received "surplus" water from the Secretary of the Interior, the watermaster of the Colorado River's Lower Basin. In the early 1990s, the six other growing Colorado River basin states pressured the Interior Department, headed by former Arizona Governor Bruce Babbitt. They wanted California to reduce its water use to its original 4.4 million acre-feet, from the 5.2 million acre-feet it had been using annually. Much of that extra water was used annually by coastal Southern California cities that were now worried about their future. River decision-makers were brought together in 1997 in Santa Fe, New Mexico, by the Water Education Foundation. I remember California Department of Water Resources Director Dave Kennedy leading informal discussions and figuring numbers literally on the back of an envelope, to see how California could reduce its water use to 4.4 million acre-feet.

By this time, agricultural to urban water transfers were seen by many as a partial solution. Concern in the Valley over potential loss of water rights spurred much debate, but in 1998, San Diego County Water Authority and IID finalized a proposed water transfer for 200,000 acre-feet of water for 75 years. It was the largest and longest voluntary water transfer agreement in the country. However, it took another agreement among the River's water-user agencies on the actual annual water use of each party, and a commitment to save the Sea, before the deal could be finalized. If the transfer happened, less water would flow into the Sea, reducing quality and quantity.

Under a 2003 Quantification Settlement Agreement (QSA), California agreed to limit its annual use to 4.4 million acre-feet. IID's Colorado River annual allotment was set at 3.1 million acre-feet. The 2003 QSA cleared the way for the San Diego and IID transfer. The transfer became the linchpin of California's plan to reduce its use of Colorado River water.

The state never fulfilled its promise to fix the Salton Sea and has done little but fund studies and small restoration projects.

The transfer, although financially lucrative, still deeply divides Valley farmers.

The deal is expected to net the district more than $2.7 billion from 2009 through 2047, and will modernize the district's system. The transfer has made the IID nearly $85 million over a 10-year period. Under the transfer plan, farmers would conserve water to free up water for the transfer. With less farm drainage, it was recognized the Salton Sea would begin to shrink. To sweeten the transfer deal, the state in the 2003 QSA negotiations agreed to use funds for a long-term fix for the Sea.

The San Diego County Water Authority and other local agencies agreed to deliver water to the Salton Sea for 15 years. In 2014, that water accounted for 10 percent of the lake's inflows. But the state never fulfilled its promise, and has done little but fund studies and small restoration projects.

Legally the agencies will stop replenishing the Sea after 2017, raising concerns that dust from the exposed seabed will contribute to respiratory problems in a region whose air quality already falls below federal standards. Debate remains on many levels, including the argument that the Sea was created by mistake, so some question spending funds to save it. A smaller but stable Sea is one possible compromise. In 2016, Memorandum of Understanding was reached between the state and federal governments to fund more habitat restoration and suppress dust at the lake in the near term. It is a start.

In a public television documentary involving the Water Education Foundation and actor Val Kilmer, who loves the Salton Sea, we noted, "It isn't just the Sea we stand to lose – it is fish, birds, farm production, and the opportunity to transfer water to growing cities, and a foundation of economic stability for the entire state." The health of the Sea is tied to the productivity of the Valley, and that in turn, ties together the people in the Valley with the people on the coast.

The story of this place is something to think about as you drive through the landscape of the Imperial Valley.

Salton Sea

God, Water and the Imperial Valley

Kevin Kelley
General Manager,
Imperial Irrigation District

The interplay between sun and water defines the ebb and flow of life in the valley and informs the psyche of the people who call it home.

For as long as I can remember, which now spans a considerable amount of time, there have been two constants in the Imperial Valley. One is the irresistible force of the sun and the other is the pervasive influence of water. Living here in what was once termed the Colorado Desert but is now charitably referred to as southernmost Southern California, it is hard to conceive of one without the other or to take either for granted.

That's because the Imperial Valley sun, which basically goes on a rampage during the summer months, can kill you, so you never want to completely turn your back on it. But the water will save your life.

It is this interplay between sun and water that defines the ebb and flow of life in the valley and informs the psyche of the people who call it home. These are people who don't have to be reminded they live in a desert, as the evidence is all around them and they couldn't forget about it if they tried. Many of them are employed in agriculture and instinctively know the food they purchase in the supermarket doesn't originate there. When they drive to work in the morning or back home at night, it isn't too many cars they must contend with, but slow-moving tractors.

Ask them about agricultural-to-urban water transfers, and they'll tell you they tend to work better on paper than on the ground, particularly where the Salton Sea is concerned. Engage them on whether the irrigation of forage crops like alfalfa hay meets the reasonable and beneficial use test, and they will ask what you have against cows (or milk). And then, if you should press them on how it is that the Imperial Valley, with more acres under cultivation than actual residents, could hold the legal right to 70 percent of the state's annual entitlement to water from the Colorado River, they will explain that's just the way God planned it.

I can't say that God played any part in bringing water to this extreme corner of the state at the turn of the last century, but I wouldn't presume to say that He didn't, either. What I do know is that everything that has transpired since then has flowed from that singular event, and there is a natural inclination, if you live here, to see the hand of providence in it. At the same time, there is a sense of foreboding that, if the drought that has gripped California persists, there will be renewed calls for the Imperial Valley to transfer more of its water, the logic being that it can do more good, for more people, somewhere else.

If droughts persist, there will be renewed calls for the Valley to transfer more of its water, the logic being that it can do more good, for more people, somewhere else.

A better course, it seems to me, would be to ensure the water transfer agreement already in place isn't undone by the state's failure to address the looming public health crisis at the Salton Sea. Just as water seeks its own equilibrium, the level of anxiety associated with it in this part of California has struggled to do the same. That anxiety, like so much else in the Imperial Valley, may be attributable to the sun.

Or maybe God really did plan it that way.

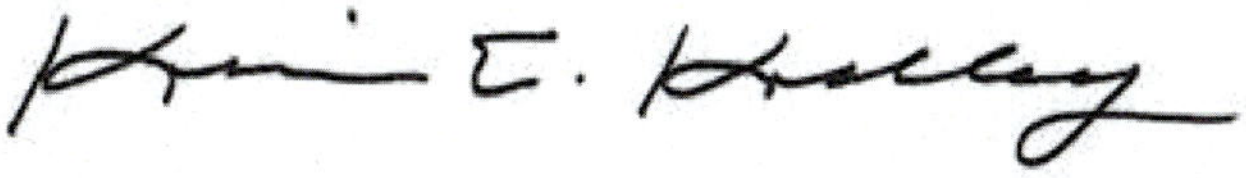

Onions in the Imperial Valley

Photo: Imperial Irrigation District

Opportunities and Challenges at the Salton Sea

Michael Cohen
Senior Associate, Pacific Institute

There are many good examples of how to maximize ecological productivity and minimize threats to public health at the Sea; we just need to mobilize political will to get it done.

I have been working to identify and implement solutions for the Salton Sea since 1998. I find the Sea's mix of ecological, hydrologic, political, and public health challenges fascinating, and often very frustrating. Unlike many other ecological challenges in the West, inflows to the Salton Sea will be measured in the hundreds of thousands of acre-feet for the foreseeable future. The challenge is how best to manage those inflows to maximize ecological productivity and minimize the threat to public health. There are many good examples of how to achieve this; we just need to mobilize the political will to get it done.

California's Salton Sea is a mass of contradictions. The Sea is actually a very salty inland lake, 50 percent saltier than the ocean yet still home to millions of tilapia, a very adaptable and resilient freshwater fish originally from Africa. The Sea is California's largest lake by surface area, stretching almost 35 miles long, but it is little more than 40 feet at its deepest. Often derided as a mistake, it's an artificial water body dependent on run-off from agricultural fields. It nevertheless, provides invaluable habitat for more than 420 species of resident and migratory birds, often numbering in the tens of thousands of individuals. The Sea lies at the very heart of a critically important water agreement that currently supplies more than 25 percent of San Diego County's water, yet it lies in a remote, arid region in southeastern California, far from the state's major cities.

The Salton Sea is the linchpin between the Colorado River and water reliability for coastal Southern California.

The nation's largest agriculture-to-urban water transfer, signed in 2003, included commitments from the state of California to backstop the costs of environmental mitigation and to undertake the restoration of the Salton Sea. Although millions of dollars have been spent on studies and meetings and staff time, there is nothing to show for it on the ground.

East side of the Salton Sea

After 2017, the transfer's impacts on the Salton Sea will no longer be offset and every drop transferred to San Diego will mean one less drop of water for the Salton Sea.

After 2017, the transfer's impacts on the Salton Sea will no longer be offset and every drop transferred to San Diego will mean one less drop of water for the Salton Sea.

The impacts will be immediate and precipitous. Salinity will triple within 10-12 years, quickly killing off the last of the fish still in the Sea, and then most of the invertebrates birds rely on for food. The surface of the lake will drop by as much as 20 feet and the shoreline will recede by as much as five miles, exposing as much as 100 square miles of dusty lakebed to the region's blowing winds, exacerbating the already poor air quality in the region and further impairing public health.

The costs of these and related impacts could run into the tens of billions of dollars within 30 years, costs that will be borne by the 650,000 residents of the Imperial and Coachella valleys.

We have known these changes were coming even before the transfer agreement was signed in 2003, but we have failed to act. The agreement provided a 15-year window to develop and implement a long-term plan for the Salton Sea, but it failed to provide annual or even periodic

benchmarks for action. One key lesson here is that future agreements of a similar scale must make continued water transfers contingent on satisfaction of clear benchmarks. This would avoid the current situation where the Imperial and Coachella valleys are implicitly deemed sacrifice zones, so that coastal Southern California can continue to grow.

While it is clearly too late to both maintain the transfer schedule and avoid catastrophic impacts to the Salton Sea, the state could still demonstrate some much-needed urgency and implement several important, achievable steps. If these steps are taken, it is not unreasonable to expect that in late 2017, state officials could host a dedication ceremony for the completion of more than a thousand acres of shallow wetland habitats at the Salton Sea and the initiation of additional air quality management and additional habitat projects. The state could also highlight the ongoing work of a Salton Sea Restoration Council plan for the next phases.

Future agreements must make continued water transfers contingent upon clear benchmarks, to avoid the current situation where these valleys are sacrifice zones so that coastal Southern California can continue to grow.

This kind of event and progress would underscore the state's commitment to protection of public and environmental health while upholding the water transfer, guaranteeing long-term water reliability for Southern California and contributing to Governor Jerry Brown's legacy of a more resilient water system and improved environment.

Wetlands at Salton Sea

Imperial Valley & the Salton Sea

Walking the Salton Sea

From June 9 to June 14, 2015, Randy Brown walked the entire circumference of the Salton Sea

Distance: 102 miles
Temperature: 110°F
Heat index: 130°F
Training: 14 months, scouting and mapping
Hazards: quicksand, mud, rivers, canals, fences, closed areas

6:02 a.m. June 9, Randy Brown starts his Salton Sea walk.

"As I walked around the shoreline of California's largest lake, the dying Salton Sea, I was constantly reminded of what a great place this lake once was. Having camped at Salton Sea dozens of times in the 1970s and 1980s as a child, I have fond memories of the lake filled with boats, fishermen, and campgrounds filled to capacity with motor homes, tents, and trailers.

Today some of the water that has kept Salton Sea alive for the last 115 years is diverted to the San Diego area. The once bustling lake is now a decrepit ghost town. It is heartbreaking to see this one-time jewel of the desert die a slow death as the pleas to save the lake seem to fall on deaf ears."

Randy Brown

Photo: Randy Brown

Some Solutions

Investment

Water Ethic

Desal

Recycling

Infrastructure

Education

Change

Environment

Stormwater

Management

Banking

Graywater

Conservation

Watersheds

Twelve Answers to California's Water Problems

Rita Schmidt Sudman

California has thrived as an 'oasis civilization,' delivering abundant water to cities and farms. Now we must adapt to aridity, punctuated by floods.

Lewis Carroll's Cheshire cat said to Alice in Wonderland, "If you don't know where you're going, any road will get you there." The message is clear. The first step to getting where you want to go is knowing where you want to end up. Next, follow a plan to get there. In the case of solving California's water problems, a strategic road map can lead to some real answers.

The water leaders of earlier generations made mistakes, but they created enough successes to give us a prosperous state. The opportunity for us is now. Many agree we have not continued the investments made by earlier generations.

Water is a shared resource. When too many users stress the water resource, it's called the "tragedy of the commons." Everyone loses. California has existed as an "oasis civilization," delivering abundant water to cities and farms. Now we must learn more about aridity, punctured by periods of unexpected floods. After four years of drought, the sky did not fall. The state is resilient and a pioneering spirit remains. In 2014, California's economy grew 27% faster than the nation's economy. Even farm earnings increased by 16% in 2015. Of course, that doesn't mean that there have not been pockets of distress and pain. It is tragic to see wells going dry, especially in poor San Joaquin Valley communities now dependent on emergency water from the state.

We've had 10 droughts in California during the last century. Some experts are calling the latest one a once-in-a-1,200 year drought exacerbated by a changing climate. Californians have responded to the Governor's call for reduced water use and dramatically cut their water use by more than the requested 25%.

Californians need water security during and beyond the current water crisis. Let's make it happen by enacting a series of changes that will take us on the road to exactly where we want to be when the next drought hits.

On the following pages are 12 answers to California's water problems. They are not my original ideas and they are not new, but this is my summary of the best solutions.

Twelve solutions to take us where we want to end up

1 Get better data so we can make better decisions

The land down under, Australia, suffered a 12-year "millennium" drought. Australians totally changed how they used water, cutting their daily use to 55 gallons a day compared to Californians' traditional use of 100 to 300 gallons a day. Australia had spent a decade before their drought requiring metering of all water withdrawals and discharges. When the drought came, it had the data to better manage the system.

California has not collected and coordinated that data, and we badly need it for analysis in both surface and groundwater. What is needed is an online database that includes information on all the surface rights. After 2009, older surface water right holders, called pre-1914 after a state water act, are now required to report every three years. The state's 2014 groundwater law and the 2015 well drillers log information law eventually will make more groundwater data available. It's a start.

None of these proposed solutions are funded at levels necessary to benefit and enrich our vibrant state.

Technology can help consumers monitor their own home water use through the use of smart meters. People can track their water online to check for water amounts in food, drinking and in their yards.

2 Both conservation and infrastructure fixes are needed

We can build more infrastructure but we cannot completely engineer our way out of the crisis. Since voter approval in 2014 of Proposition 1, a $7.5 billion water bond measure, the construction of new dams is more probable as $2.7 billion in funding was set aside for water storage projects. There are five key sites under consideration by the California Water Commission in the Sacramento and San Joaquin valleys.

Although there probably will be some more large reservoirs built in California, these dams will be different from earlier dams. Due to climate change, consideration should be made for earlier warm rain filling the reservoir rather than snowpack. By formula, all major reservoirs also must free up space for winter flood flows, even if the year ultimately turns out to be dry and the reservoirs don't fill up. Better use also could be made of the state's groundwater basins for future storage. These basins hold at least three times as much usable water as the surface reservoirs.

Since the drought of the early 1990s, California's cities and urban water agencies have developed aggressive conservation programs. By the height of the drought in 2015, many of these agencies beat Governor Brown's conservation mandate, saving an average of 31% compared to the previous two years ago. Incentives for drought tolerant landscapes have worked. Some water agencies are paying for each square foot of lawn replaced by drought-tolerant landscaping.

Saving one square foot of grass in Southern California saves 42 gallons of water a year, according to the Metropolitan Water District of Southern California (MWD), which pays $2 for each square foot of lawn replaced by drought-tolerant landscaping. This cash for grass program has been so popular that MWD ran short of funds in 2015, having received more than $330 million in applications for rebates. On the upside, changing our plant palate to match our Mediterranean climate is gaining popularity and someday rebates may not be needed.

The downside for the water agencies is that they lose money when water sales fall, due to strong conservation. In 2015, revenues fell by about $500 million. Eventually agencies must increase rates to make up the shortfall as many agency costs remain fixed. The court ruling on Proposition 218, a constitutional amendment affecting water pricing, has caused problems when the agencies need to set higher rates and send a conservation signal to high water users. We need to find a legal way to overcome a recent court ruling not allowing this price message to consumers.

Farmers and agricultural water districts also have made investments in irrigation efficiency particularly since the congressional reforms in 1992.

3 Aggressively implement the 2014 groundwater law

Surface water cutbacks have hit San Joaquin Valley farmers hard and they have turned to groundwater to grow the higher value crops that produce better revenue. It was estimated in 2014, California farmers substituted groundwater for over 75 percent of the irrigation water they did not get from surface systems. Farmers are increasing pumping by an estimated one billion gallons a day. That is not a long-term sustainable plan and they know it.

The 2014 Sustainable Groundwater Management Act, passed to reduce overdraft, does not start implementation until 2020, although there is evidence that banks are changing their long-term farm lending practices based on the law. The Public Policy Institute of California (PPIC) says this is a sign that the market may help quicken the pace of the law's implementation. If intense groundwater pumping continues, there may be other

Tomatoes grown with recycled water

methods needed in the short term, including mitigation fees and county emergency ordinances that restrict new or deeper wells in areas of overdraft. Surface water and groundwater are one resource and must be managed together.

4 Gutting environmental laws will not solve the problem

It may come as a shock that much of the impact of the drought has fallen on the fish. There are few tools for environmental managers to improve the situation. According to the PPIC, 18 native fish species are at a high risk of extinction, including most runs of salmon and several species of trout and a number of other fish. The cause is loss of spawning habitat from reduced flows and increased water temperatures. In the future, drastic measures have been suggested to shift current hatcheries from producing fish for commercial and recreation fisheries and into ones that attempt to save species.

A key to the fishery problem is find a better method to convey water through California's Delta, the area where much of the state's water converges and the place where Delta pumps destroy fish. It's a dilemma: water must be conveyed in some way through the Delta, both to meet the legal commitments of sending some water south, and to create enough flow for migrating salmon. There have been many proposed solutions to physically fix the situation. Although the state and some stakeholders are moving ahead with a plan for conveyance, solutions are controversial. While fish are being pulled in the wrong direction – sucked toward the Delta pumps – environmental problems that cause the pumps to be shut down periodically will continue. Some type of infrastructure fix is necessary.

5 Change public perception in order to increase investment in recycled water

Remember the hydrologic water cycle you learned about in 4th grade? Recycled water is water that is used more than one time before it passes into the natural system. Since civilization began, people have recycled water. Today water from the Colorado River goes through six states on its way to California. It is taken from the river, used, treated and returned to the river by 200 wastewater plants before it is disinfected for the last user downstream in Southern California.

By not dumping billions of gallons of treated sewage into the ocean, recycled water could yield up to 1.1 million acre-feet of water annually, enough to supply 8 million people – one-fifth of the state's population.

Currently, the vast majority of our recycled water is dumped in rivers and the ocean. We could do more water recycling if we could get over a perception it is water that goes from "toilet to tap." In fact, there is no toilet to tap. We put cleaned-up water into groundwater or surface reservoirs, environmental buffers, before additionally disinfecting it and then drinking it. In that way, we are mimicking nature's hydrologic cycle. This recycled water has been called the single largest source of water supply in the state for the next 25 years. Other countries recycle on a big scale. Israel treats 86% of its domestic wastewater and recycles it for about half of the country's agricultural use.

By not dumping hundreds of billions of gallons of treated sewage into the Pacific Ocean, experts estimate recycled water could yield up to 1.1 million acre-feet of water annually, enough to supply 8 million people – one-fifth of the state's population.

Southern California has led the way in recycling for years. Decades ago, Orange County Water District pioneered pumping the treated water into an aquifer and holding it for six months before it becomes drinking water. MWD is partnering with wastewater agencies to create recycling projects. In Northern California, San Francisco in 2015 passed an ordinance requiring new buildings of a certain size have on-site recycling systems that reuse their own wastewater. Even in the San Joaquin Valley city of Modesto, a city where water has been abundant and inexpensive, and a place where the historic arch over the main street says "Water, Wealth, Contentment, Health," money now is being spent to recycle water rather than discharging that water into the San Joaquin River. The reason is that recycled water is dependable and drought-proof. Consumers are lining up for free recycled water from local agencies for use on yards. Of course, there may be unintended consequences of putting less sewage water into the pipes to move solids along, but these factors can be handled with options, including building fewer sewage treatment plants in the future.

6 *Use urban stormwater in our own yards*

We need new ways to use water falling on our homes and yards. Before cities urbanized, rain

Drought tolerant landscape uses stormwater drainage in Palm Springs.

would seep into the earth and return to groundwater. Since runoff could cause destructive local flooding in urban and suburban areas, planners devised rain gutters and underground storm drains to quickly move the water away. The runoff picks up speed and pollutants as it runs over streets, sidewalks and parking lots and is lost to future use. We cannot afford this waste.

When it rains in Los Angeles, up to 10 billion gallons of water can surge into storm drains and flow straight in the Pacific Ocean. The L. A. Department of Water and Power hopes to build three large projects to collect stormwater and put it into the groundwater basin for recharge and later use. They could collect up to 10 times more stormwater each year by 2035. This is an aggressive program and it will not be inexpensive. It also will depend on continued cleanup of the basin, contaminated in some locations from past practices of industries and consumers.

Meanwhile, consumers can get involved by using permeable driveways and other surfaces that let water soak into groundwater basins. Another way to collect stormwater is rainwater harvesting – not just a utopian idea anymore. Rain gardens and cisterns can hold thousands of gallons of water and can also be installed in residential properties, as done in Arizona. In Australia more than a quarter of city residents installed cisterns during their long drought.

7 Use technology to increase the use of graywater in our homes

Since 2010 the state has encouraged use of graywater (also called grey water). Building codes caught up with handy Californians who were hooking up their own graywater systems. Now laundry to landscape systems can be installed without a permit. Water savings are significant. A University of California study found there could be enough graywater in Southern California to water all the residential landscapes. The approved systems require water lines to run beneath the soil or mulch so the water does not come in contact with people. The water can even be used on vegetable gardens as long as it doesn't touch root crops like carrots. Top builders in California now offer in-home water recycling.

One of the state's largest builders, KB Homes, has inaugurated a new on-site water recycling system that recycles water from the home to irrigate landscapes. This system is a standard feature in one of their San Diego county projects. It moves graywater from showers, bathtubs, washing machines and bathroom sinks (toilet, kitchen sink and dish washer water is "black water") and then passes the water through filters removing most solids, impurities and bacteria. The water goes into a 200-gallon underground tank that feeds the home's landscape irrigation system. Imagine taking a shower without guilt because your shower is watering your flowers!

The San Diego County Water Authority says that the typical water use of 161 gallons per person per day could be cut in half by this type of recycled system, combined with high-efficiency home appliance features. The catch is that the system adds $10,000 to the new house price and about $15,000 to build into an existing home. But given the cost of landscaping and preserving it, some homeowners will consider the price for water security worth it. This type of system could be expanded and will drop in price as technology and production continues.

"Why do diamonds cost more than water?"
Adam Smith

8 Value water more accurately

In college Econ 101, I remember the classic economic paradox example about the price of water and its value. Adam Smith posed the question: Why do diamonds cost more than water? We were taught that even though water is necessary for our survival, our economy puts prices on items based on scarcity and value. Since water was plentiful, it remained cheap—at least compared to diamonds.

In a drought-stressed society, the question is: Is water too cheap? Currently, the price of water in our homes is much less than we usually pay for cell phone service or cable television. Without price signals on water, there is no financial incentive for the consumer to conserve and use water wisely. Using price to reduce consumption should be acceptable. In Israel, a two-tiered price system was introduced in 2009. Those who consume more than the basic allotment pay a higher rate. If in California we could legally use a tiered price system, we would see more conservation. To protect low-income consumers, the first block of water—about 15 gallons a day necessary for survival—could be subsidized by those paying higher rates.

9 Increase water marketing

Water markets are an important part of the solution to combat scarcity. This has existed on a sporadic basis for 25 years, but critics say it's difficult to get approvals to market water and the system needs refining and streamlining. Infrastructure systems need to be improved and connected to provide flexibility to move water from willing sellers to buyers. Many observers support easing more regulations to allow those with the older, more valuable water rights to lease or sell their water. The older rights are often held by farmers and some are willing to shift their water use for the right price. Taken to the extreme, water marketing could negatively affect local communities and take water from the environment and downstream users. But handled with care, it's a valuable way to free up water for many uses.

The severe drought has caused scholars and others to take a look at the question of a complete makeover of California water law. It would take decades of court challenges to make water rights changes. A 1978 Commission to Reform Water Rights Law came out of the 1977 drought with major suggestions that were never enacted. Incentives like water marketing are a better way to effect change.

10 Use desalination where appropriate

Seven years ago, a drought in Israel nearly led to dry pipes in homes. Today there is an abundance of water because five major desalination ("desal") plants went into production. More than a dozen California coastal communities are looking at desal to supply a portion of their water needs. A $1 billion project in Carlsbad is providing 50 million gallons of drinking water a day for San Diego County residents, about 7% of the counties water supply. It is a costly proposition.

To remove the salt from seawater is 50 percent of the cost. It takes 38 megawatts, enough to power 28,500 homes to push 100 million gallons of brackish water a day through a series of reverse osmosis filters. During its mega drought, Australia built half a dozen big desal plants. Those in Brisbane, Sydney and Melbourne are on standby incurring costs but not being used as the rains have returned.

A desal plant can provide drought resiliency if the costs can be blended into other water costs. That's the case in Carlsbad where the San Diego County Water Authority is blending the cost over county users. The Carlsbad plant took six years of permitting, and 14 lawsuits and appeals to be built. Industry advocates say the entire future of desal is riding on this project. The price tag is high but, because the area gets only 10 inches of rain a year, and has no significant groundwater basins to store water, it may be worth the price to ease reliance on imported water. The Water Authority will pay about $2,000 an acre-foot to buy the water from Poseidon, the private company building the project. And the average customer's bill will increase only $5 to $7 a month because of the blended system. For communities with limited water supply options, it can be a good insurance policy.

11 Better coordination and some consolidation among water agencies and utilities

The State Water Board estimates there are 3,000 water providers, 1,100 wastewater organizations, 600 irrigation districts, 140 reclamation districts and 60 flood control agencies in California. They don't often coordinate. Certainly wastewater, water supply and stormwater agencies could work more closely toward a more holistic use of water. Also, if a loss of funding continues, there may be additional motivation for some districts to consolidate. There definitely is a need to collaborate and go beyond agency limits. Governor Brown proposes giving the State Water Board power to order the consolidation of water agencies, if the board concludes it would improve residents' water supply. Funds would go to pay for equipment and connection work and would come from Proposition 1, passed by the voters in 2014.

The Santa Ana Watershed Project Authority (SAWPA) is a group of water agencies operating under a "one water one watershed concept" from the forest source to the sea. With the strength of agencies working together, SAWPA is able to get funding from bond and loan programs allowing for larger development of water resources than they could get individually. From developing forest management policies to funding desal to recharging groundwater basins, SAWPA is promoting the health of the entire watershed. It's a good example of agencies working together and a model for others to embrace.

Knowledge brings power to solve problems and can put us on the road to where we want to end up.

12 Imagination, innovation and a new water ethic are needed.

With water shortages, people look for someone to blame. Who's to blame?

Is it the farmers and ranchers, and their use of water for the products they grow and produce?

Is it the people in the cities and suburbs with their lawns and landscapes?

How about the industrial community?

How much water does it take to make a movie, a computer chip or a beer?

Are the environmentalists at fault because laws protect endangered fish and the fishing industry?

There's nothing like a crisis to focus thinking and political will. And there will be more critical droughts and floods in California's future. Yet, California can continue to lead the nation in ingenuity, imagination and innovation. We can use those resources. We also need to subscribe to a new water ethic, a conscious awareness of our water use and sustainability. We all need to know where our water comes from, how much we use, what we put in it, and where it ends up after we use it.

Knowledge brings power to solve problems and can put us, like Alice, on the road to where we want to end up.

Embracing Thirty Years of Evolution in Water Management in California

Tim Quinn
Executive Director,
Association of California
Water Agencies

We have done a complete 180 degree turn in California water strategy from the generations that preceded us.

For those of us who live and breathe California water, these are fascinating times. Though some headlines would have you believe our state is on the precipice of collapse, that is far from the case. Yes, we are in a millennial drought, but the resiliency we've built into the system over the past 30 years – particularly on the local level – has allowed us to stay strong throughout this crisis. I believe it will continue to do so as we move forward, but only if we continue to invest in a statewide comprehensive plan for the future.

When I look back on my 30 years in the water world, I see tremendous change in how we manage water. The modus operandi of the 20th century was to build huge projects to move vast amounts of water to support a growing economy. These efforts spawned major projects such as the Los Angeles Aqueduct, Hetch Hetchy Reservoir, Pardee Reservoir, and the Colorado River Aqueduct. Now, the story you hear from water managers is very different. They are investing in local resources – recycling, desalination, cleaning up contaminated groundwater and capturing storm water. We have done a complete 180 degree turn in California water strategy from the generations that preceded us, and this is what is protecting us today during drought. We are relying on new technologies and new strategies. We are a changed industry.

The 1987-1992 drought wasn't as severe hydrologically as the most recent, but we are better prepared now than we were a quarter century ago. Today, urban areas are undergoing rationing to reduce "ornamental" water use, but the core urban economies remain untouched, by and large, because urban water managers invested in storage and other tools and reduced demand. We've knocked down per capita water use and will continue to do so. We can greatly reduce water use on our green lawns and lush median strips without shutting down businesses.

Not so with agriculture. The farm economy is more impacted because they don't have these ornamental uses of water. Their water goes directly to crops which are their businesses. If the drought continues into year 5, 6 and 7, then we may cut into the economic uses of water in both agricultural and urban areas. That should happen through the market, not regulation.

> *It is good to know that a comprehensive plan is on the table. Now we need to continue to implement it.*

I admit that I am an optimist and I am proud to be a part of these powerful changes. At the Association of California Water Agencies (ACWA), we've carefully and strategically chosen areas where we can make a difference. We've established policy principles on re-thinking storage for the 21st century, headwaters protection and groundwater sustainability. Some of these principles – particularly groundwater – have been adopted into law. We also developed a Statewide Water Action Plan to look at water comprehensively. Important elements of this strategy have been embraced by the Brown Administration. As I approach the end of my career, it is good to know that a comprehensive plan is on the table. Now, we need to continue to implement it.

I came of age as a teenager in the John F. Kennedy era that placed high value on public service. Making a difference is important to me. Water has turned out to be the most fascinating public policy career I ever could have imagined. I've seen enormous progress over the past 30 years in water and am absolutely sure that California will stay on that path of progress.

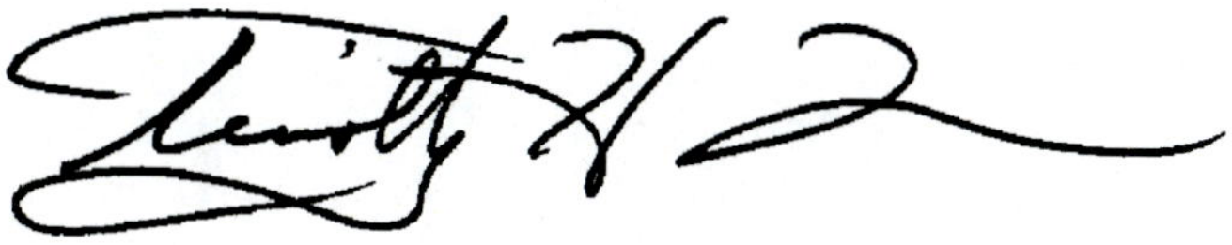

Some Solutions

Sustaining a Healthy Environment

Kim Delfino
California Program Director,
Defenders of Wildlife

Rarely do people say that they do not want to protect the environment. Our choices are seldom consistent with these values.

When I was young, each summer my father used to load up our old Jeep with fishing gear and we would spend days wandering back roads in the Sierra Nevada in search of the perfect fishing spot. Years later, influenced by those summers in the Sierra, I attended law school with the desire to defend our environment, particularly fish and wildlife. It did not take long before I started to work on water issues, as nothing can exist without water. During my time at Defenders of Wildlife, I have come to work on two seemingly unconnected issues: Central Valley wetlands and the Salton Sea.

As California has developed, we have destroyed more than 90 percent of California's wetlands. This means that the protection and restoration of wetlands along the Pacific Flyway from the Salton Sea to the Central Valley are critical to the survival of hundreds of species of birds, millions of fish, and a multitude of reptiles, amphibians, insects, and mammals.

I started working on Central Valley wetland and Salton Sea issues naively. I thought that they could be fixed with money and better enforcement of our laws. I was wrong. Solving our water dilemmas are about more. The solutions are about what society values and how we choose to protect those values. Rarely do people say that they do not want to protect the environment. However, our choices are seldom consistent with these values. Or, when we finally choose conservation, it is in reaction to a greater problem that could have been avoided if we had acted wisely earlier, even if it might have been a difficult solution to secure.

The current drought has demonstrated the risk with this approach. We have strained our fish and wildlife resources by managing them on the razor's edge. Delta smelt trawl numbers have come back with the stomach-dropping number of zero, wild salmon numbers are precariously low, and there are fewer wetlands upon which birds can rest and feed when they migrate. This pattern of choices has created an unsustainable future for California's fish and wildlife unless we change our approach.

The Salton Sea is no different. As part of a massive water transfer in which water, previously used to irrigate fields in the Imperial Valley and then allowed to drain into the Salton Sea, was diverted to urban Southern California. The Sea was given a 15-year reprieve in which a “solution” was to be developed. If nothing is done, the Sea faces an ecological and public health disaster of epic proportions, with choking dust storms caused by the exposed sea bed. It would be a Sea devoid of life due to high salinity levels, and millions of displaced birds – nearly two-thirds of all migrating species -- that relied on the Sea as a critical stopover.

With two years left before the mitigation water ceases to be delivered to the Sea and no “solution” yet to be chosen, the Sea is running of time.

I believe we can solve these problems, but it will take a recognition by all parties that protection of our natural resources is a key societal value – not just a talking point -- and a willingness to make hard choices before we push our environment past the point of no return.

We have strained our fish and wildlife resources by managing them on the razor’s edge.

Pope Francis recently issued a groundbreaking encyclical on the environment, “Laudato Si,” in which he eloquently explains why care of the environment is a requirement, not a recommendation. He details why protection of our natural world is a fundamental value that we must make room for in our lives. If we follow that advice in our water policy choices and make room for sustaining healthy fish and wildlife resources, we will create a healthier California for future generations of people, fish and wildlife.

Knowledge is Power When it Comes to Water

Jennifer Bowles
Executive Director,
Water Education Foundation

I've seen the power and value that comes with people knowing more about water - where their water comes from and where it goes.

I count myself among the water geeks. When I studied water law in Colorado during a year-long fellowship at Colorado University, I hiked to the headwaters of the Colorado River in Rocky Mountain National Park in heavy snow, just to get a look at the area. The drive up Highway 395 from Los Angeles past the Los Angeles Aqueduct in the Owens Valley, is always a thrill, as the landscape provides a virtual history lesson in California water. When I lived in Riverside, I biked along the Santa Ana River as it ebbed and flowed through the seasons.

Now living in Sacramento as executive director of the Water Education Foundation, I bike along the American River, a robust and dynamic waterway. I've backpacked along the Tuolumne River and into Hetch Hetchy Valley, curious to understand why John Muir launched such a ferocious battle against San Francisco's plan to dam the river and bring that water to the city.

It was natural that when I started my journalism career as an Associated Press general assignment reporter in Southern California, I gravitated to the water stories. I quickly saw that knowledge of the physical and political water system could help me to cover this issue more effectively. Water has a steep learning curve for reporters – for anyone – from understanding the complicated plumbing that brings the water to our faucets, to the water quality issues that are often measured in parts-per-billion. As a reporter, I was searching for clear, unbiased information. The Water Education Foundation became my lifeline.

It's been said that all politics is local. I certainly saw that when I became a water reporter at the Press-Enterprise in Riverside. I had to delve into the politics of local plumbing. And, as most of us realize, the plumbing is often at the heart of issues in the water world.

This is an exciting time to be involved in water issues. The drought, though painful, has spurred new groundwater management laws, more science efforts, increased funding

The Foundation's Project WET (Water Education for Teachers)

for recycling, stormwater and graywater projects, and remote sensing for understanding water supply and use. At the Foundation, we are following discussions on climate change, questioning our current water rights system and learning how new technologies can solve problems.

> *Water has a steep learning curve for anyone, from understanding complicated plumbing to water quality issues measured in parts-per-billion.*

None of these changes would have happened without the current public spotlight on water. I am fully dedicated to helping the public understand more about water so they can make better decisions. I don't think this is a Pollyannaish idea. I've seen the power and value that comes from people understanding water - where our water comes from and where it goes after we use it. Whether it's a vote for a local water board member or someone in state or federal office, we need to know their position on water issues. We are all invested personally in our water system. If we are asked to conserve water and change our lifestyles, we will need knowledge and commitment to make changes.

While I don't think many people will become water geeks like me, I'm glad I started down this path many years ago. Like my predecessor at the Foundation, I'm passionate about education. From where I started as a water reporter, with my role at the Foundation I'm excited that I can contribute by bringing diverse interests together to seek and find solutions to California's water issues.

Jennifer

California Water Timeline 1769 - 2016

1769 First permanent Spanish settlements established in San Diego. Water rights established by Spanish law.

1848 Gold discovered on the American River. Treaty of Guadalupe signed, California ceded from Mexico, California Republic established.

1850 California admitted to Union. Construction begins on Delta levees and channels.

1860 Legislature authorizes the formation of levee and reclamation districts.

1862 Major flood in Sacramento Valley inundates new city.

1880 First flood control plan for the Sacramento Valley developed by State Engineer William Hammond Hall.

1884 Federal Circuit Court decision in Woodruff v. North Bloomfield requires termination of hydraulic mining debris discharges into California rivers.

1886 California Supreme Court decision in Lux v. Haggin reaffirms legal preeminence of riparian rights, upheld again 40 years later.

1892 Conservationist John Muir founds Sierra Club.

1901 First California deliveries from the Colorado River made to farmland in the Imperial Valley.

1902 U.S. Bureau of Reclamation established by the Reclamation Act of 1902.

1905 First bond issue for the city of Los Angeles' Owens Valley project; second bond issue in 1907 approved for actual construction.

Colorado River flooding diverts the river into Imperial Valley, forming the Salton Sea.

1908 City of San Francisco's filings for Hetch Hetchy project approved.

1913 Los Angeles Aqueduct begins service.

1920 Col. Robert B. Marshall of the U.S. Geological Survey proposes statewide plan for water conveyance and storage.

1922 Colorado River Compact appropriates 7.5 million acre-feet per year to each of the river's two basins.

1923 Hetch Hetchy Valley flooded to produce water supply for San Francisco despite years of protest by John Muir and other conservationists.

East Bay Municipal Utility District formed.

1928 Congress passes Boulder Canyon Act, authorizing construction of Boulder (Hoover) Dam and other Colorado River facilities.

Federal government assumes most costs of the Sacramento Valley Flood Control System with passage of the Rivers and Harbors Act.

California Constitution amended to require that all water use be "reasonable and beneficial."

St. Francis Dam collapses, flooding the Santa Clarita Valley, killing more than 450 people.

Worst drought of the 20th century begins; ends in 1934, establishing benchmark for water project storage and transfer capacity of all major water projects.

1931 State Water Plan published, outlining utilization of water resources on a statewide basis.

County of Origin Law passed, guaranteeing counties the right to reclaim water from an exporter if it is needed in the area of origin.

1933 Central Valley Project (CVP) Act passed.

1934 Construction starts on All-American Canal in the Imperial Valley (first deliveries in 1941) and on Parker Dam on the Colorado River.

1937 Rivers and Harbors Act authorizes construction of initial CVP features by U.S. Army Corps of Engineers.

1940 MWD of Southern California's Colorado River Aqueduct completed; first deliveries in 1941.

1944 Mexican-U.S. Treaty guarantees Mexico 1.5 million acre-feet per year from Colorado River.

1945 State Water Resources Control Board created.

1951 State authorizes the Feather River Project Act (later to become the State Water Project).

First deliveries from Shasta Dam to the San Joaquin Valley.

1955 Flood in the Sacramento Valley kills 38 people.

1957 California Water Plan published.

1959 Delta Protection Act resolves some issues of legal boundaries, salinity control and water export.

1960 Burns-Porter Act ratified by voters; $1.75 million bond issue to assist statewide water development.

1963 Arizona v. California lawsuit decided by U.S. Supreme Court, allocating 2.8 million acre-feet of Colorado River water per year to Arizona.

1964 Partially completed Oroville Dam helps save Sacramento Valley from flooding.

1966 Construction begins on federal New Melones Dam on the Stanislaus River after 20 years of controversy over the reservoir's size and environmental impacts; completed in 1978.

1968 Congress authorizes Central Arizona Project (CAP) to deliver 1.5 million acre-feet of Colorado River water a year to Arizona.

Congress passes Wild and Scenic Rivers Act.

1970 Passage of the National Environmental Quality Act, California Environmental Quality Act and California Endangered Species Act.

1972 California Legislature passes Wild and Scenic Rivers Act preserving the north coast's remaining free-flowing rivers from development.

Federal Clean Water Act (CWA) passed.

1973 First SWP deliveries to Southern California.

1974 Congress passes Safe Drinking Water Act.

1978 State Board issues Water Rights Decision 1485 setting Delta water quality standards.

1980 State-designated wild and scenic rivers placed under federal Act's protection.

1982 Proposition 9, the Peripheral Canal package, overwhelmingly defeated in statewide vote.

Reclamation Reform Act raises from 160 acres to 960 acres the amount of land a farmer can own and still receive low-cost federal water.

1983 California Supreme Court in National Audubon Society v. Superior Court rules that the public trust doctrine applies to Los Angeles' diversion from tributary streams of Mono Lake.

Dead and deformed waterfowl discovered at Kesterson Reservoir in western San Joaquin Valley, pointing to problems of selenium-tainted agricultural drainage water.

1986 Ruling by State Court of Appeals (Racanelli Decision) directs State Board to consider all beneficial uses, including instream needs, of Delta water when setting water quality standards.

Passage of Safe Drinking Water and Toxic Enforcement Act (Proposition 65) prohibiting discharge of toxic chemicals into state waters.

Coordinated Operation Agreement for CVP and SWP operations in the Delta signed.

1987 State Board's Bay-Delta Proceedings begin to revise D-1485 water quality standards.

1989 In a separate challenge to Los Angeles' Mono Basin water rights, an appellate court holds that fish are a public trust resource in California Trout v. State Water Resources Control Board.

MWD and Imperial Irrigation District agree that MWD will pay for agricultural water conservation projects and receive the water conserved.

1991 MOU signed to implement urban water conservation programs.

Inyo County and Los Angeles agree to jointly manage Owens Valley water, ending 19 years of litigation.

West Coast's first municipal sea water desalination plant opens on Catalina Island.

1992 Congress approves landmark CVP Improvement Act.

1993 Federal court rules in Natural Resources Defense Council v. Patterson that the CVP must conform with state law requiring release of flows for fishery preservation below dams.

Arizona's CAP declared complete by the federal government.

1994 State Board amends Los Angeles' water rights licenses to Mono Lake.

Bay-Delta Accord sets interim Delta water quality.

CALFED Bay-Delta Program planning initiated.

1995 State Board adopts new water quality plan for the Delta and begins hearings on water rights.

1997 New Year's storms cause state's second most devastating flood of the century.

SWP's Santa Barbara Aqueduct completed.

1999 Sacramento splittail minnow and spring-run Chinook salmon added to federal endangered species list.

2000 CALFED Record of Decision signed by state and federal agencies.

2001 Klamath Project irrigation water crisis.

2002 Voters approve Proposition 50, a $3.44 billion bond issue to fund improvements in water quality and reliability.

2003 Interior Secretary orders California's Colorado River allocation limited to 4.4 million acre-feet; water users sign Quantification Settlement Agreement (QSA).

Paterno v. State of California ruling by third Appellate District Count determined the state of California liable for potentially hundreds of millions of dollars because state accepted levee without any measures to ensure it met design standards and then failed in 1986 flood.

SWP contractors, DWR and environmental groups settle lawsuit over the Monterey Amendment.

2004 State Board initiates review of 1995 Water Quality Control Plan.

Congress approves long-awaited legislation to re-authorize CALFED.

2005 Scientific surveys of the Delta and Suisun Marsh reveal ongoing, sweeping population crash of native pelagic fish.

Legislation directs DWR to evaluate the future of the Delta.

2006 Legal settlement among a coalition of environmental and fishing groups, the U.S. Departments of the Interior and Commerce and the Friant Water Users Authority ending a 18 year legal battle over claims to release water from Friant Dam to maintain a live stream for fish to the Merced River.

2007 SWP pumping operations shut down to protect endangered Delta smelt (Wanger Decision).

DWR estimates that Delta levees are vulnerable to massive failure if major earthquake occurs.

Seven Colorado River states agree to new drought rules and shortage criteria.

Delta Vision Blue Ribbon Task Force releases plan. Other Delta planning processes continue.

2008 DWR initiates Bay Delta Conservation Plan (BDCP) EIS/EIR.

Governor declares statewide drought after second dry/critical year.

2009 U.S. District Judge Oliver Wanger rules federal government did not analyze impact of Delta smelt protection rules on water exporters. Gov. Schwarzenegger signs a comprehensive water package designed to achieve Delta co-equal goals.

San Joaquin River Restoration Act passed by Congress.

USGS report finds that about 60 million acre-feet of groundwater has been lost in the San Joaquin Valley since 1961.

2010 Judge Wanger rules the federal government failed to consider impacts to water users when it restricted pumping from Delta to protect Chinook spring-run salmon and other fish.

Final Klamath Basin Restoration Agreement and Klamath Hydroelectric settlement Agreement signed. Implementation contingent on authorizing legislation, funding and environmental review.

2012 Central Valley Flood Protection Plan adopted by state flood board.

Bureau of Reclamation releases Colorado River Basin Water Supply and Demand study that projects a range of future water supply and demand imbalances for the seven Colorado River states.

2012 Monumental five year agreement, Minute 319, sets stage for improved Colorado River water supply reliability between the United States and Mexico.

The state legislature approves a bill to put a water bond on the 2014 ballot.

2013 State Water Resources Control Board continues update of flow objectives.

Final chapters of Bay Delta Conservation Plan (BDCP) released proposing long-term Delta restoration and promoting a more reliable statewide water supply through the creation of twin tunnels to move water beneath the Delta to the existing state and federal pumping facilities.

2014 Gov. Jerry Brown signs legislation creating local agencies to oversee groundwater pumping, making California the last state in the West to regulate groundwater.

Brown signs a $7.5 billion water bond that will go before California voters. The bond contains $2.7 billion for new storage, $1.49 billion for watershed restoration, $810 million for water reliability projects, $520 million for water quality projects, $725 million for water recycling, $900 million for groundwater cleanup and $395 million for flood management.

California voters approve the water bond with about 67 percent of the vote.

2015 In the midst of a four-year drought, Gov. Jerry Brown orders first-ever, statewide water reductions aimed at urban California.

Gov. Brown unveils California EcoRestore, a reduced Delta habitat conservation plan (from 100,000 to 30,000 acres) that does not seek a 50 year permit to take water from Delta, as proposed under the former BDCP plan. Governor reiterates support for California WaterFix, the new name for proposed twin tunnels conveyance system.

2016 Gov. Brown releases his administration's program of ten purposed actions with investment in diverse programs to create a more resilient state water system by 2019.

State Water Board adopts rules to require all water diverters to measure and report how much water they use annually.

President Obama signs controversial water act that shifts 25 years of California federal policy on environmental water and sends more water through Delta, streamlines potential dam construction while providing minor funding, funds desalination and recycling projects and maximizes water supplies for the CVP and SWP users.

2017 State Water Board expected to rule midyear to mandate more water flows through the Delta to Pacific Ocean. About 30% - 50% less water from major Northern rivers would be available to be pumped to farms and on to Southern California.

Timeline Courtesy of Water Education Foundation

Glossary of Water Words

acre-foot – 325,851 gallons, or enough to cover a football field to a depth of one foot. An average California household uses between one-half and one acre-foot of water per year.

anadromous fish – fish species, such as salmon, that migrate from fresh water streams to the ocean and back to complete their life cycle.

appropriative rights – water right based on physical control over water, or based on a permit or license for its beneficial use.

area of origin – water right statutes began in 1931 to protect local areas against export of water. These statues have seldom been invoked.

conjunctive use – the planned use and storage of surface and groundwater supplies to improve water supply reliability.

desalination — the process of removing salt and minerals from seawater and from brackish water, to create drinkable water.

developed water – water that is controlled and managed (dammed, pumped, diverted, stored in reservoirs or channeled in aqueducts) for a variety of uses.

graywater (also called greywater) is all wastewater generated in households or office buildings from streams without fecal contamination. Sources of graywater include sinks, showers, baths, clothes washing machines and dishwashers.

groundwater – water that has seeped beneath the earth's surface and is stored in the pores and spaces between alluvial materials (sand, gravel or clay).

hydrologic cycle – movement of water as it evaporates from rivers, lakes or oceans, returns to earth as precipitation, flows into rivers to the ocean and evaporates again.

instream uses – the beneficial uses of water within a river or stream, such as providing habitat for aquatic life, sport fishing, river rafting or scenic beauty.

public trust doctrine – doctrine rooted in Roman law which holds that certain natural resources are the property of all, to be held in trust by the state.

recycled water – the treatment and reuse of wastewater to produce water of suitable quality for additional use.

riparian right – a water right based on the ownership of land bordering a river or waterway.

stormwater - rainwater that falls on rooftops, collects on driveways, roads and sidewalks and is carried away through a system of pipes that is separate from the sewerage system.

surface water – water that remains on the earth's surface, in rivers, streams, lakes, reservoirs or oceans.

Glossary Courtesy of Water Education Foundation

Author Biographies

Stephanie and Rita in 2012, with Stephanie's painting created for the Water Education Foundation in 1997.

Rita Schmidt Sudman

Rita Schmidt Sudman served as Executive Director of the respected nonprofit Water Education Foundation for 34 years. She fostered resolution of water resource issues in California and states sharing the Colorado River through facilitation, education and outreach. She directed the development of publications, public television programs, tours, press briefings and school programs. Since retiring in 2014, she serves as the Foundation's Senior Advisor.

A former radio and television reporter and producer in San Francisco and San Diego, she received her master's degree in telecommunications from San Diego State University. Her television production team at the Foundation won three regional Emmys, several regional Emmy nominations and one Telly award for their public television documentaries.

She has served on numerous boards including the University of California President's Advisory Commission on Water and the national board of Water For People, an international program for water in developing countries. She has received awards from major agricultural, environmental and urban water groups including the Lifetime Achievement Award from the California Groundwater Resources Association and the Service to the Water Industry Award from the American Water Works Association. She was named a "Superhero" by the California State Fair for her water education work.

Stephanie Taylor

Artist and writer Stephanie Taylor is inspired by curiosity about place and by how people relate to each other. Her work calls upon whatever materials are best suited to sites--from paint, steel, and cement to words.

Taylor has led a multifaceted life as a professional artist, first in Los Angeles (where she started as an award winning advertising art director and earned a degree in history from UCLA) and since 1983, in Sacramento. She has created over 30 exterior mural and sculpture projects in Northern California, and made site-specific installations, sculpture and paintings for major institutions and collections all over the United States as well as in Paris and Kyoto. Her work has been exhibited in many fine art galleries.

While earning a masters in sculpture in 2006, Taylor discovered a renewed interest in writing--especially about California. Since 2011, the Sacramento Bee Opinion has published her paintings and essays about people and the environment as the series "California Sketches." Taylor is a licensed artist in Disney's fine art program and has published a small book of photography, Keyholes in Tuscany. Articles about her work have appeared in many publications, from the Sacramento Business Journal, Los Angeles Times to international design magazines. She has been interviewed about her work and art issues on NPR's Capital Public Radio.

Acknowledgements

Rita Schmidt Sudman

Most of all I want to thank Stephanie Taylor for convincing me to author this book with her. After my retirement, she saw that I had something more to contribute from my years covering water issues. With Stephanie's beautiful art and interesting essays, this book is a much richer product than the water policy book I might have written.

My background in water came largely from my work leading the Water Education Foundation. The journey that I started there 35 years ago led to my total involvement and concern about this fascinating topic. I also appreciate the able work of my successor, Jennifer Bowles, to continue the Foundation's work in educating the public. The Foundation remains a tremendous source of information about California and Colorado River issues.

Finally, I can't thank enough my husband, John Sudman, for his patience throughout the year-long project and his insightful comments and editing. I also appreciate the efforts of our daughter, Suzanne Sudman, who works in the film industry, to help me make sure the water information could be understood by non-water types. Suzanne grew up learning about water and being taken, whether she liked it or not, to water locations all over California and the West.

Stephanie Taylor

First, of course, I'd like to thank Rita Schmidt Sudman for her support and enthusiasm, and patience with me not allowing her to "retire." Gratitude to Gary Reed, Editorial Board member and Forum editor of the Sacramento Bee, who took a leap of faith when an artist declared she wanted to write for him; "California Sketches" allows me to wander California in search of answers to satisfy my endless curiosity. Thanks to all the many scientists, farmers and others who allowed me to interrupt busy days, and to Dr. Eldridge Moores who gave me one of my favorite quotes. Thanks to the Squaw Valley Community of Writers who inspired me as a participant and as a party crasher. To my writing mentor, Rae Gouirand, who is one of the smartest people I know, and my family and dear friends who tolerate my explorations, endless appreciation.

Stephanie and Rita

We want to thank the 21 diverse water experts who contributed original pieces to this book. Rita asked people she especially respected – regardless of whether or not we agreed with their points of view – to write short essays describing why they continue to be involved in this field, what they are passionate about and what solutions they suggest for the future. To our delight, they submitted some very intriguing pieces.

We also want to give special thanks to Cindy Nickles who carefully proof read and edited the book. Cindy has worked for more than 20 years in the water industry and we appreciate her careful editing and knowledge on California water issues. Thanks also to Penny Hill who is a journalist and writer and who formerly worked on projects for the Water Education Foundation. Her suggestions and edits provided clarity and unity to our book. We also appreciate the assistance of two longtime friends and writers, Judy Maben and Pat McArdle. Judy Maben formerly worked at the Water Education Foundation and has kept current on water issues. Patricia McArdle, author of Farishta, is a former U.S. State Department diplomat who saw water struggles first-hand around the world. Also helpful as editors on specific articles were Karen Saunders and Julie Nygaard.

Finding the right photo is often difficult and time consuming. Stephanie and I appreciate the photos we received from government agencies. Thanks go to Lisa Navarro at the Bureau of Reclamation and to Mike Miller, Bill Kelley and Ted Thomas at California Department of Water Resources. Other photo credits are listed separately.

44692405R00138

Made in the USA
San Bernardino, CA
20 January 2017